An Illustrated Guide to the Benthic Marine Algae of Coastal North Carolina

I. Rhodophyta

An Illustrated Guide to the Benthic Marine Algae of Coastal North Carolina

I. Rhodophyta

Donald F. Kapraun

The University of North Carolina Press
Chapel Hill

© 1980 The University of North Carolina Press
All rights reserved
Manufactured in the United States of America
ISBN 0-8078-4063-7
Library of Congress Catalog Card Number 79-21566

Library of Congress Cataloging in Publication Data

Kapraun, Donald F.
    An illustrated guide to the benthic marine algae of coastal North Carolina.

Bibliography: v. 1, p.
Includes index.
CONTENTS: 1. Rhodophyta.
1. Marine algae--North Carolina--Identification.
2. Coastal flora--North Carolina--Identification.
I. Title. II. Title: Benthic marine algae of coastal North Carolina.
QK571.5.N8K36   589'.39'2   79-21566
ISBN 0-8078-4063-7 (v.1)

Contents

Acknowledgments vii

Introduction 3

   Location and Description of the Area 6

   Nature and Floristic Affinity of the Red Algae 7

Systematic List of Coastal Red Algae 11

Key to the Species 17

Systematic Account 33

Literature Cited 87

Illustrations 101

Glossary of Selected Terms 197

Index 203

Acknowledgments

The author is indebted to the encouragement and guidance provided by Dr. Max Hommersand over the several years required to formulate the scope of this project and to solve certain taxonomic problems. Drs. R. B. Searles and C. Schneider critically read the manuscript, offered much helpful advice, and generously provided information from their work in offshore North Carolina waters. Thanks are also due Ed Flynn, who provided many of the initial collections, and Paul Gabrielson for critically examining specimens of the Solieriaceae. This project was supported by The University of North Carolina at Wilmington Marine Science Program.

An Illustrated Guide to the Benthic Marine Algae of Coastal North Carolina

I. Rhodophyta

Introduction

Historically, studies of the marine algae of Atlantic North America have been centered in New England (Harvey 1852; Collins 1903, 1909; Farlow 1881), the tropical shores of south Florida (Harvey 1852; Taylor 1928) and offshore islands (Collins and Hervey 1917; Børgesen 1913-20; Howe 1918, 1920). Taylor's classic survey (1957, 1960) of the Atlantic coast consciously reflected this bias by emphasizing the disparate nature of these two floras and treating the Carolinas as a transitional area which seasonally included cool temperate and tropical species. Hoyt's study (1920), the exception to this general neglect of the mid-Atlantic coast, was unfortunately limited in that collections were made only during the warmer months and almost exclusively in the vicinity of Beaufort, North Carolina.

In modern times, numerous workers have provided information on the inshore flora of the Carolinas (Blomquist and Humm 1946; Williams 1948a, 1949; Aziz and Humm 1962; Earle and Humm 1964; Brauner 1975; Richardson 1979) and the extensive flora of the offshore reefs and continental shelf (Williams 1948b, 1951; Searles 1972; Schneider and Searles 1973, 1975, 1976; Schneider 1974, 1975a, 1975b, 1976). Also, taxonomic studies have begun to appear (Fiore 1969, 1977; Rhyne 1973; Aregood 1975; Schneider 1975c; Coll and Cox 1977; Kapraun 1977a), but no attempt has been made to unify this information into a regional flora. Because so many additions to our flora were made in checklists or ecological notes, and at a time when the extreme morphological variation inherent in some taxa was not appreciated, doubtless cases of synonymy have occurred, and species have been attributed to our shores on the basis of incorrectly identified specimens. According to Dixon (1963), it is doubtful if more than 70% of the species listed in any flora anywhere can be identified accurately as a consequence of poorly delimited taxa and synonyms being uncritically conserved by successive authors. Assuming that the studies in North Carolina have not been uniquely accurate, a large number of the 176 species of red algae so far reported for the state probably cannot be confirmed (Searles and Schneider 1978). A more realistic total of 135 to 140 can be extrapolated by comparing the 79 species encountered in this study with the 104 red algal species reported offshore (Schneider 1976).

A necessary first step in the development of a flora comparable to those now available for the eastern North Atlantic (Ardré 1970a, 1970b; Dixon and Irvine 1977; Gayral 1958, 1966; Kornmann and Sahling 1977; and Rueness 1977) is a published account based on adequate collections, critically described and illustrated, and complemented by culture studies. A reliance solely on previously published accounts and herbarium specimens lacking morphological or reproductive features essential for critical determinations cannot but perpetuate past errors and frustrate any attempt to achieve an accurate understanding of our local flora.

    The present study of the coastal red algae and a volume treating both green and brown algae which is to follow, are not presumed to be all inclusive, even though a conscientious attempt was made to collect seasonally in most of the state's varied habitats (Fig. 1). Rather, these guides will serve their purpose if they are viewed as a summary of littoral species present during an intensive six-year study, as well as one view of the morphological ranges to be accommodated by specific taxa. In addition, it is hoped that they will call attention to certain problem groups, e.g., Acrochaetiaceae and Solieriaceae, which have not been satisfactorily treated to date and should be critically investigated. Finally, it is hoped that these illustrated seaweed guides will provide students with a means of identifying and better understanding our local marine algae. It is unfortunate that no such resource is currently available. Hoyt's paper (1920) is both out of date and out of print. Taylor's books (1957, 1960) are largely too technical for introductory students, while of necessity treating cursorily the Carolinas, which are on the fringe of both the northeast cool temperate and tropical Caribbean floras that are their intended subjects.

    The format of this guide was planned with students in mind. The key is strictly dichotomous with paired alternatives chosen representing anatomical features which are typically present and easily observed. Technical terminology is held to a minimum and those terms deemed necessary are explained in the glossary and illustrated in one or more drawings. Additional definitions and illustration of terms useful in the identification of marine algae are now available elsewhere (Hine 1976; Abbott and Hollenberg 1976; Woelkerling 1976). Finally, the descriptions are based on North Carolina specimens collected during this study. This is especially valuable because species with cool temperate or tropical floristic affinities which occur in the Carolinas often vary considerably from the descriptions available for them elsewhere (Taylor 1957, 1960).

The study of marine algae involves specialized collection and preservation techniques. Several texts (Taylor 1960; Dawson 1966a; Dawes 1974) and regional guides (Edwards 1970; Scagel 1967; Dawson 1966b) provide detailed instructions for beginning students and will not be repeated here.

LOCATION AND DESCRIPTION OF THE AREA

Algae included in this guide were collected primarily in southeastern North Carolina (Fig. 1). A review of the pertinent literature on the geology and hydrology of this general area was given by Schneider (1976). Since nearly the entire study area consists of barrier islands of fine, unconsolidated sand fronting a system of sounds, bays, and estuaries, collections were typically restricted to man-made jetties which provide stable substrate in the littoral necessary for benthic algal growth. The exceptions to this general rule are grassbeds *(Zostera marina)* in the shallow bays near Beaufort (Brauner 1975), behind Topsail Island, and at Sneads Ferry in the New River estuary; the exposed littoral coquina outcroppings near Kure Beach; and the consolidated sediment outcroppings which form a broken ridge in shallow water (10-15 m) within 2-3 km of Masonboro Island and Wrightsville Beach. Eulittoral oyster reefs in tidal creeks and estuaries provide additional habitats and in winter often support surprisingly rich algal growths.

NATURE AND FLORISTIC AFFINITY OF THE RED ALGAE

The Atlantic coast from Cape Hatteras, North Carolina to Cape Kennedy, Florida, characterised by minimum winter sea temperatures of 9-14°C and summer maxima of 26-29°C, is considered a warm temperate biogeographic region (Stephenson and Stephenson 1952; van den Hoek 1975). Since the Carolinas have a relatively low degree of endemism compared with the warm temperate Mediterranean-Atlantic region (van den Hoek 1975), the area seems to be essentially a transition zone between the better-developed cool temperate and tropical floras (Humm 1969). Indeed, of the 79 species of red algae treated in this study, 26 (34%) are reported for New England (Taylor 1957), while 61 (78%) are reported for the American tropics (Taylor 1960; Børgesen 1913-20), and only four littoral species seem to be endemic to the Southeast: *Porphyra carolinensis*, *P. rosengurtii*, *Branchioglossum minutum*, and *Calonitophyllum medium*. However, recent studies on the genus *Polysiphonia* in North Carolina have provided evidence that some species with warm temperate-tropical distributions are more accurately described as having a warm temperate floristic affinity (Kapraun 1977a, 1977c). Likewise, a critical investigation of the 29 species (37%) in this study also reported from warm temperate western Europe (Gayral 1966; Ardré

1970a; Feldmann-Mazoyer 1941) would probably demonstrate that many have a warm temperate floristic affinity.

One of the most striking features of the littoral algal flora in North Carolina is its pronounced seasonality (Hoyt 1920; Williams 1948a, 1949; Brauner 1975; Kapraun 1977a). During the warmer months, eurythermal tropical species of Carribbean affinity predominate, while in winter species of cool temperate New England affinity achieve their maximum growth and reproductive development. Superimposed over these two extreme patterns of seasonality is a third, less conspicuous assemblage consisting of warm temperate and cosmopolitan species adapted to warm temperate conditions (Stephenson and Stephenson 1952). These species occur throughout the year, but tend to achieve maximum development in spring and fall (Kapraun 1978c). Culture studies have shown these patterns of seasonal periodicity to be primarily temperature-related phenomena (Edwards 1969, 1970; Kapraun 1977b; Fiore 1977; Richardson 1979), but light intensity and photo-period have also been implicated (Edwards 1969, 1971; Kapraun 1978a).

In addition to seasonal cycles, more subtle changes in species composition and abundance from year to year are often reported in regional studies (Conover 1958; Phillips 1961; Edwards and Kapraun 1973; Kapraun 1974). In our area, alternating northerly and southerly water currents are probably responsible for introducing stenothermal cool temperate and tropical species, respectively. For example, during July and August, pelagic *Sargassum natans* and *S. fluitans* wash ashore in great abundance along the coast of Onslow Bay. These masses of brown algae support a luxuriant flora of epiphytes (Woelkerling 1972a, 1975), including *Polysiphonia howei*, *P. flaccidissima*, *P. pseudovillum*, and *Anotrichium barbata*. The two latter species are previously unreported from our coast and, presumably being more stenothermal, fail to become permanently established.

Schneider (1976) reported that the benthic algal flora on the continental shelf of the Carolinas, which may represent the only truly subtropical biogeographic region in Atlantic North America, is more strongly tropical than the reported littoral flora. He attributed this permanent northward displacement of species with a Caribbean affinity into temperate latitudes to the mild seasonal bottom temperatures created by the nearby Gulf Stream. During this present study, several species previously reported only from deep water were found in southern Onslow Bay in the littoral, e.g., *Branchioglossum minutum*, or at moderate depth (to 10 m), e.g., *Halymenia floridana* and *Gracilaria blodgettii*, suggesting water temperatures there to be more moderate than in the Beaufort region.

Systematic List of Coastal Red Algae

SUBCLASS BANGIOPHYCEAE

ORDER GONIOTRICHALES
    FAMILY GONIOTRICHACEAE
        Goniotrichum alsidii (Zanard.) Howe

ORDER BANGIALES
    FAMILY ERYTHROPELTIDACEAE
        Erythrocladia subintegra Rosenv.
        Erythrotrichia carnea (Dillw.) J. Ag.
        Erythrotrichia ciliaris (Carm. ex Harv.) Thur. in Le Jol.
    FAMILY BANGIACEAE
        Bangia atropurpurea (Roth) C. Ag. (= *Bangia fuscopurpurea* [Dillw.] Lyngb.)
        Porphyra carolinensis Coll et Cox
        Porphyra rosengurtii Coll et Cox

SUBCLASS FLORIDEOPHYCEAE

ORDER NEMALIALES

FAMILY ACROCHAETIACEAE

Audouinella alariae (Johnsson) Woelk. (= *Kylinia alariae* [Johnsson] Kylin)

Audouinella dasyae (Coll.) Woelk. (= *Acrochaetium dasyae* Coll.)

Audouinella hallandica (Kylin) Woelk. (= *Acrochaetium sargassi* Børg.)

Audouinella microscopica (Naeg. in Kuetz.) Woelk. (= *Kylinia crassipes* [Børg.] Kylin)

Audouinella secundata (Lyngb.) Dix. (= *Colaconema secundata* [Lyngb.] Woelk.)

Audouinella thuretii (Bornet) Woelk. (= *Acrochaetium thuretii* [Bornet] Coll. et Herv.)

FAMILY CHAETANGIACEAE

Scinaia complanata (Coll.) Cott.

ORDER GELIDIALES

FAMILY GELIDIACEAE

Gelidium crinale (Turn.) Lamour.

Gelidium americanum (Tayl.) Santelices (= *Pterocladia americana* Tayl.)

ORDER CRYPTONEMIALES

FAMILY SQUAMARIACEAE

Peyssonnelia rubra (Grev.) J. Ag.

FAMILY CORALLINACEAE

Dermatolithon pustulatum (Lamour.) Fosl. (= *Lithophyllum pustulatum* [Lamour.] Fosl.)

Heteroderma lejolisii (Rosanoff) Fosl. (= *Fosliella lejolisii* [Rosanoff] Howe)

Amphiroa beauvoisii Lamour.

Corallina cubensis (Mont.) Kuetz.

Jania adhaerens Lamour.

FAMILY GRATELOUPIACEAE

Grateloupia filicina (Wulf.) C. Ag.

Halymenia floridana J. Ag.

Halymenia gelinaria Coll. et Howe

ORDER GIGARTINALES

FAMILY GRACILARIACEAE

Gracilaria blodgettii Harv.

Gracilaria foliifera (Forssk.) Børg.

Gracilaria sjoestedtii (Kylin) Papenf. (= *Gracilariopsis sjoestedtii* [Kylin] Dawson)

Gracilaria verrucosa (Huds.) Papenf.

FAMILY SOLIERIACEAE

Meristotheca floridana Kylin

Neoagardhiella baileyi (Harv. ex Kuetz.) Wynne et Tayl. = *Agardhiella baileyi* [Harv. ex Kuetz.] Tayl.)

FAMILY HYPNEACEAE

Hypnea cornuta (Lamour.) J. Ag.

Hypnea musciformis (Wulf.) Lamour.

FAMILY PHYLLOPHORACEAE

Gymnogongrus griffithsiae (Turn.) Mart.

FAMILY GIGARTINACEAE

Gigartina acicularis (Wulf.) Lamour.

ORDER RHODYMENIALES

FAMILY RHODYMENIACEAE

Rhodymenia pseudopalmata (Lamour.) Silva

Botryocladia occidentalis (Børg.) Kylin

FAMILY CHAMPIACEAE

    Champia parvula (C. Ag.) Harv.

    Lomentaria baileyana (Harv.) Farl.

ORDER CERAMIALES

    FAMILY CERAMIACEAE

        Anotrichium barbatum (C. Ag.) Naeg. (= *Griffithsia barbata* C. Ag.)

        Anotrichium tenue (C. Ag.) Naeg. (= *Griffithsia tenuis* C. Ag.)

        Antithamnion cruciatum (C. Ag.) Naeg.

        Callithamnion byssoides Arn. ex. Harv. in Hook.

        Ceramium byssoideum Harv.

        Ceramium diaphanum (Roth) Harv.

        Ceramium fastigiatum Harv. in Hook.

        Ceramium rubrum (Huds.) C. Ag.

        Griffithsia globulifera Harv.

        Spermothamnion investiens (Crouan Frat.) Vick.

        Spyridia hypnoides (Bory) Papenf. (= *Spyridia aculeata* [Schimp.] Kuetz.)

    FAMILY DELESSERIACEAE

        Branchioglossum minutum C. Schn.

        Caloglossa leprieurii (Mont.) J. Ag.

        Calonitophyllum medium (Hoyt) Aregood (= *Hymenena media* [Hoyt] Tayl.)

        Grinnellia americana (C. Ag.) Harv.

        Hypoglossum tenuifolium (Harv.) J. Ag. var. *carolinianum* L. Williams

FAMILY DASYACEAE

Dasya baillouviana (Gmel.) Mont. (= *Dasya pedicellata* [C. Ag.] C. Ag.)

FAMILY RHODOMELACEAE

Bostrychia radicans (Mont.) Mont. (= *Bostrychia rivularis* Harv.)

Chondria dasyphylla (Woodw.) C. Ag.

Chondria littoralis Harv.

Chondria polyrhiza Coll. et Herv.

Chondria tenuissima (Good. et Woodw.) C. Ag.

Herposiphonia tenella (C. Ag.) Ambr. (= *Herposiphonia secunda* [C. Ag.] Ambr.)

Laurencia corallopsis (Mont.) Howe

Laurencia poitei (Lamour.) Howe

Micropeuce mucronata (Schm.) Kylin (= *Brongniartella mucronata* [Harv.] Schm.)

Polysiphonia denudata (Dillw.) Kuetz.

Polysiphonia ferulacea Suhr

Polysiphonia flaccidissima Hollenb.

Polysiphonia harveyi Bailey

Polysiphonia havanensis Mont. sensu Børg.

Polysiphonia howei Hollenb.

Polysiphonia macrocarpa Harv.

Polysiphonia nigrescens (Huds.) Grev.

Polysiphonia pseudovillum Hollenb.

Polysiphonia sphaerocarpa Børg.

Polysiphonia subtilissima Mont.

Polysiphonia tepida Hollenb.

Polysiphonia urceolata (Lightf.) Grev.

Pterosiphonia pennata (Roth) Falkenb.

Key to the Species

1. Erect plants essentially a uniseriate (Fig. 6) branched or unbranched filament, but may become pluriseriate (Fig. 4) in older parts; microscopic or but a few cm in length   2
1. Plants crustose or structurally more complex than a uniseriate filament; if filamentous, then showing evidence of pericentral cells   17
2. Plant unbranched   3
2. Plant branched   5
3. Plant a solitary uniseriate microscopic epiphyte
    Erythrotrichia carnea (page 34)
3. Plant longer, becoming pluriseriate, forming dense epiphytic tufts or epilithic mats in the littoral fringe   4
4. Plants epiphytic, rose red; cells subquadrate to elongate
    Erythrotrichia ciliaris (page 34)
4. Plants epilithic, purple black; cells several times broader than tall
    Bangia atropurpurea (page 35)

5. Cells with a single chromatophore; plants minute
    epiphytes less than 1 cm tall   6
5. Cells with many small chromatophores; plants larger,
    more than 1 cm tall   12
6. Branching dichotomous; cells red, squat, embedded
    in a thick matrix
    Goniotrichum alsidii (page 33)
6. Branching alternate or unilateral; reproduction in
    part by lateral or terminal monospores   7
7. Basal system unicellular or with a few accessory
    cells   10
7. Basal system multicellular: filamentous or pseudo-
    parenchymatous   8
8. Basal system of extensive branching filaments;
    erect filaments often with elongate hairlike
    terminal cells
    Audouinella thuretii (page 41)
8. Basal system limited, consisting of a multicellular
    pad or a few short accessory filaments   9
9. Basal system parenchymatous (Fig. 25); lateral
    branches sometimes bearing unicellular hairs;
    plants to 2 mm tall
    Audouinella secundata (page 41)
9. Basal system consisting of original spore which
    becomes obscured by short accessory filaments;
    plants to 1 mm tall
    Audouinella dasyae (page 40)
10. Basal cell spherical or swollen; branching secund
    or arcuate (Fig. 15)   11

Key to the Species : 19

10. Basal cell cylindrical; branching alternate
    Audouinella hallandica (page 41)
11. Plants minute, about 100 μm tall; branching simple, arcuate; monosporangia sessile (Fig. 21)
    Audouinella microscopica (page 41)
11. Plants larger, to 1 mm tall; branching secund; monosporangia mostly stalked
    Audouinella alariae (page 40)
12. Plants irregularly branched; every cell of the main axis not bearing a branch   13
12. Plants regularly branched; nearly every cell of the main axis bearing a branch   16
13. Plants erect to 6 cm tall; cells large, swollen, barrel-shaped and bearing whorls of dichotomously branched trichoblasts (cf. Fig. 101)
    Griffithsia globulifera (page 64)
13. Plants small and more delicate; erect branches to 3 cm tall, arising from decumbent filaments; cells cylindrical   14
14. Plants light pink, delicate and soft, forming dense epiphytic tufts; erect filaments to 6 mm tall; cells uninucleate, 40-60 μm diameter, not constricted at the nodes
    Spermothamnion investiens (page 64)
14. Plants rose red, gregarious; erect filaments rigid, 2-3 cm tall; cells multinucleate, 50-120 μm diameter noticeably constricted at the nodes   15
15. Trichoblasts scarce or lacking; cells to 70 μm diameter
    Anotrichium tenue (page 59)

15. Apical cell and younger filaments bearing conspicuous thrice-branched trichoblasts; cells to 120 μm diameter
    Anotrichium barbatum (page 59)
16. Branching alternate, plumose (Fig. 109)
    Callithamnion byssoides (page 61)
16. Branching opposite or whorled
    Antithamnion cruciatum (page 60)
17. Plants erect and calcified, or crustose　　　　　　　　18
17. Plants not calcified; structurally more complex than crustose plants with obvious erect blades or branching thalli　　　　　　　　24
18. Plants crustose　　　　　　　　19
18. Plants erect, highly branched　　　　　　　　22
19. Plants not calcified, minute, epiphytic on coarse algae
    Erythrocladia subintegra (page 34)
19. Plants more or less calcified, epiphytic or epilithic　　　　　　　　20
20. Plants pink to dark red, forming epilithic crusts 2-3 cm diameter
    Peyssonnelia rubra (page 45)
20. Plants pink to white, forming crusts to 5mm diameter, epiphytic on sea grasses or coarse algae　　　　　　　　21
21. Vegetative crusts monostromatic (Fig. 37); adjacent thalli confluent
    Dermatolithon pustulatum (page 47)
21. Vegetative crusts 1-3 layers of vertically elongated cells; adjacent thalli remain distinct
    Heteroderma lejolisii (page 47)

Key to the Species : 21

22. .Segments to 1 mm broad and compressed; conceptacles scattered over their surface
    Amphiroa beauvoisii (page 47)
22. Segments cylindrical and slender, less than 500 μm diameter; conceptacles marginal or terminal 23
23. Branching dichotomous
    Jania adhaerens (page 48)
23. Branching opposite
    Corallina cubensis (page 48)
24. Plants membranous 25
24. Plants not membranous; if foliaceous, then gelatinous 32
25. Plants with a conspicuous midrib 26
25. Plants without a midrib, but veinlets may be present ... 29
26. Plants with long lanceolate blades, to 25 cm tall
    Grinnellia americana (page 68)
26. Plants minute, less than a few cm long 27
27. Branching dichotomous from constricted nodes
    Caloglossa leprieurii (page 67)
27. Branching from the midrib or margins 28
28. Branching marginal
    Branchioglossum minutum (page 66)
28. Branching from the midrib
    Hypoglossum tenuifolium (page 69)
29. Plants dark red to purple; blades simple, broadly lanceolate, delicately thin 30
29. Plants pink to rose; blades more or less dichotomously branched, divisions narrow, less than a few cm wide 31

30. Plants less than 4 cm long, ovate to lanceolate; marginal teeth present
    Porphyra carolinensis (page 36)
30. Plants to 30 cm long, oblong, ovate, or linear lanceolate; margin entire
    Porphyra rosengurtii (page 37)
31. Branching irregularly dichotomous, veinlets numerous, margins undulate, plants delicately membranous
    Calonitophyllum medium (page 67)
31. Branching dichotomous, no veinlets visible, margins entire, plants firm
    Rhodymenia pseudopalmata (page 56)
32. Plants broadly foliaceous, gelatinous to membranous   33
32. Plants otherwise: cylindrical, filamentous, etc.   35
33. Medulla of vertical pillarlike filaments; blades large, to 25 cm tall, ovate, very gelatinous, 300-500 µm thick
    Halymenia gelinaria (page 49)
33. Medulla of irregularly branched anastomosing filaments; blades membranous   34
34. Plants small, 4-10 cm tall, becoming palmate, 100-150 µm thick; medulla of filaments of various sizes, cortex of one-to-few small cells (Fig. 51)
    Halymenia floridana (page 49)
34. Plants larger, to 25 cm tall, marginally branching, 300-400 µm thick; medulla of thin filaments below a subcortex of large, irregular anastomosing cells (Fig. 68)
    Meristotheca floridana (page 53)

Key to the Species : 23

35. Plants hollow throughout or in part, or consisting
     of slender axes supporting slime sacs   36
35. Plants otherwise: cylindrical, but not hollow,
     filamentous, etc.   39
36. Plants of wiry axes supporting vesicular slime sacs
         Botryocladia occidentalis (page 57)
36. Plants hollow except for internal partitions (septa),
     or persistent multiaxial medullary strands   37
37. Plants entirely hollow, but with internal parti-
     tions; branching irregularly alternate or secund   38
37. Plants hollow, but with persistent multiaxial
     medullary filaments; branching dichotomous
         Scinaia complanata (page 42)
38. Plants with secund branches constricted and
     septate at their base
         Lomentaria baileyana (page 58)
38. Plants divided into barrel-shaped hollow segments
     separated by partitions
         Champia parvula (page 57)
39. Plants wiry or bushy, growing in pulvinate tufts or
     forming a dense turf   40
39. Plants otherwise: cartilaginous, gelatinous, or
     filamentous, but not forming wiry tufts   42
40. Plants bushy, repeatedly forking; medulla of angular
     cells
         Gymnogongrus griffithsiae (page 55)
40. Plants wiry, axes of small diameter; rhizines in
     subcortical area   41

41. Plants small, 1-2 cm tall, axes flattened, sparingly pinnately branched; medulla of a few parallel cell rows, rhizines scarce
    Gelidium americanum (page 45)
41. Plants taller, to 10 cm tall, axes terete, irregularly branched; medulla of numerous parallel cell rows intermixed with abundant rhizines
    Gelidium crinale (page 44)
42. Plants cartilaginous to firm or gelatinous, cylindrical or with flattened branches; uniaxial or multiaxial   43
42. Plants clearly filamentous or structurally comprised of an axial cell row more or less covered with pericentral cells   57
43. Plants slender, terete, covered with numerous radially placed spinelike branchlets (Fig. 75)   44
43. Plants coarser, cylindrical or flattened, or, if small, with marginally pinnate branching; not covered with spinelike branchlets   45
44. Plants large and bushy, to 20 cm tall; some branches with thickened crozier tips (Fig. 76); branchlets simple and spinelike
    Hypnea musciformis (page 55)
44. Plants smaller with one-to-few main axes, to 10 cm tall; crozier tips lacking; branchlets crowded, pedicellate and stellate (Fig. 80)
    Hypnea cornuta (page 55)
45. Transection showing a filamentous medulla, at least in part   46

Key to the Species : 25

45. Transection showing a parenchymatous medulla of
large cells 47
46. Plants less than 5 cm tall, flattened, branching
marginally pinnate
Grateloupia filicina (page 48)
46. Plants to 30 cm tall, terete, with radial branching
of several orders
Neoagardhiella baileyi (page 53)
47. Plants cylindrical, with short, blunt-tipped branchlets terminated by whorls of trichoblasts 48
47. Plants cylindrical or with flattened branchlets;
branches resembling main axes or much more slender;
if terete, then tapering to a point 53
48. Lateral branchlets short, blunt, not basally constricted 52
48. Lateral branchlets elongate, constricted at the
base, and spindle-shaped 49
49. Plants minute, creeping, to 5 mm tall
Chondria polyrhiza (page 72)
49. Plants 10-30 cm tall, erect from a discoidal base 50
50. Branchlets club-shaped, constricted at the base,
apices blunt with sunken apical cell
Chondria dasyphylla (page 71)
50. Branchlets spindle-shaped, apical cell exposed 51
51. Branchlets solitary, basally constricted, but not
tapering to both ends; terminal trichoblasts
conspicuous
Chondria littoralis (page 71)

51. Branchlets often in dense clusters, tapering to both ends; exposed apical cell conspicuous with few surrounding trichoblasts
    Chondria tenuissima (page 72)
52. Plants rose red, branching subdichotomous from near the base, branches blunt-tipped and resembling main axis; branchlets few and inconspicuous
    Laurencia corallopsis (page 74)
52. Plants to 20 cm tall, yellow brown, cartilaginous, becoming entangled; branchlets peglike, to 2 mm long
    Laurencia poitei (page 74)
53. Plants olive green, entangled in cushions, 3-8 cm tall, fibrous axes repent; in transection showing a cortex of long, anticlinally branched cell rows (Fig. 66)
    Gigartina acicularis (page 56)
53. Plants attached, usually with several main axes, to 30 cm tall or more, if entangled; showing in transection a cortex of 2-3 layers of small cells   54
54. Main axes extremely long to 50 cm, little branched; showing in transection a cortex of 4-5 cells in rows; cystocarps with highly filamentous gonimoblast (Fig. 60)
    Gracilaria sjoestedtii (page 51)
54. Main axes shorter, usually less than 30 cm tall, abundantly branched; showing in transection a cortex of 2-3 cell layers; cystocarps with parenchymatous gonimoblast   55
55. Plants noticeably flattened, at least at branch axils; branches often 1 cm or more wide
    Gracilaria foliifera (page 50)

Key to the Species : 27

55. Plants bushy, terete throughout, branches less than
5 mm diameter 56
56. Plants rose colored; branchlets distinct from main
axes, constricted at their bases; tetrasporangia
limited to branchlets
  Gracilaria blodgettii (page 50)
56. Plants olive to black, coarse and firm, often
becoming free; branchlets repeatedly dividing,
ultimately ending in fine tips; tetrasporangia
scattered over main axes and branches
  Gracilaria verrucosa (page 52)
57. Plants mostly less than 15 cm long; setaceous
branchlets lacking 60
57. Plants 15 cm or more long; bushy with short, setaceous branchlets (Fig. 141) 58
58. Main axes lacking pericentral cells, but corticated
with a layer of alternating bands of long and short
rectangular cells (Fig. 138)
  Spyridia hypnoides (page 65)
58. Main axis with 5 pericentral cells more or less
corticated by rhizoidal filaments 59
59. Plants 15-20 cm tall; delicately branched, axes
less than 1 mm diameter, ramelli to 1 mm long
  Micropeuce mucronata (page 75)
59. Plants to 50 cm tall; with branches resembling main
axes, 4-6 mm diameter; heavily setaceous with
slender ramelli 4-6 mm long
  Dasya baillouviana (page 69)

60. Central axial cells corticated by nodal rings of smaller cells which may spread to cover the internodes, but axial row remains visible at least at the tip ... 61
60. Central axial cells completely covered by polysiphonous cells (Fig. 227) ... 64
61. Plants large and bushy, to 25 cm tall; central axial cells completely covered by a corticating layer of small cells except at the branch tips (Fig. 134)
    Ceramium rubrum (page 63)
61. Plants smaller; central axial cells naked; corticating cells restricted to nodal bands (Fig. 117) ... 62
62. Plants small, less than 2 cm tall; creeping; corticating band of 2 cell rows or with lower cells distinctly broader than tall ... 63
62. Plants larger, to 10 cm tall; erect; nodes ultimately with 6 or more rows of cells; adventitious branching abundant in older parts (Fig. 119)
    Ceramium diaphanum (page 62)
63. Corticating band of one-to-few equal-sized cells
    Ceramium fastigiatum (page 63)
63. Corticating band of 4 or more cell rows with the lower 1-2 cells distinctly broader than tall
    Ceramium byssoideum (page 61)
64. Plants with distichously pinnate branching (Fig. 227) ... 65
64. Plants creeping or with radial branching ... 66
65. Filament tips monosiphonous; main axis with 6-8 pericentral cells
    Bostrychia radicans (page 70)

65. Filament tips polysiphonous; main axis with 8-9 pericentral cells
    Pterosiphonia pennata (page 84)
66. Plants essentially erect, branching radial, or, if prostrate branches present, their apices not inrolled   67
66. Plants often epiphytic, essentially creeping with horizontal apices inrolled
    Herposiphonia tenella (page 73)
67. Pericentral cells 5 or more   76
67. Pericentral cells 4   68
68. Plants erect from a discoidal base, large and stout; branching alternate from a distinct main axis 400-500 µm diameter; some basal cortication present
    Polysiphonia harveyi (page 78)
68. Plants with a base of creeping filaments or some prostrate branches; main axis less than 250 µm; uncorticated   69
69. Branches arising in axils of trichoblasts; erect axes from an extensive creeping base   70
69. Branches developmentally replacing trichoblasts   71
70. Plants to 10 cm tall; unicellular rhizoids remain in open connection with pericentral cells (Fig. 201)
    Polysiphonia havanensis (page 78)
70. Plants minute epiphytes; unicellular rhizoids cut off from proximal end of pericentral cells
    Polysiphonia flaccidissima (page 77)
71. Rhizoids cut off from pericentral cells   74
71. Rhizoids remain in open connection with pericentral cells   72

72. Plants minute, 1-2 cm tall, black; erect filaments
little branched
Polysiphonia macrocarpa (page 80)
72. Plants larger, 4-10 cm tall, purple to light brown;
erect filaments highly branched   73
73. Plants to 5 cm tall; erect filaments dichotomously
branched; found in brackish, muddy water
Polysiphonia subtilissima (page 82)
73. Plants to 10 cm tall; erect filaments with numerous
short, alternate divergent branches; found in
exposed coastal habitats
Polisyphonia urceolata (page 84)
74. Plants minute, to 2 mm tall, with an extensive
creeping system
Polysiphonia pseudovillum (page 81)
74. Plants taller, 0.5 cm or more, lacking a creeping
system but with few decumbent branches (Fig. 192)   75
75. Plants small, 0.5-2.0 cm tall, often epiphytic;
axes 60-200 µm diameter
Polysiphonia sphaerocarpa (page 81)
75. Plants large, 5-10 cm tall, 150-250 µm diameter;
trichoblasts abundant and conspicuous
Polysiphonia ferulacea (page 76)
76. Pericentral cells 5-6, plants erect to 15 cm tall
Polysiphonia denudata (page 76)
76. Pericentral cells 7 or more   77
77. Pericentral cells 12 (10-14); plants large, to 15 cm
tall, coarse, dark red to black
Polysiphonia nigrescens (page 80)
77. Pericentral cells 7-10; plants creeping, delicate   78

78. Plants 1-2 cm (to 10 cm) tall; pericentral cells 7-8; branches arising in axils of trichoblasts
  Polysiphonia tepida (page 83)
78. Plants minute epiphytes; pericentral cells 8-10 branches replacing trichoblasts
  Polysiphonia howei (page 79)

Systematic Account

SUBCLASS BANGIOPHYCEAE

ORDER GONIOTRICHALES

FAMILY GONIOTRICHACEAE

Goniotrichum alsidii (Zanard.) Howe   Fig. 3

Plants microscopic epiphytes consisting of branched uniseriate filaments, pink to rose; cells with a single stellate plastid and embedded in a thick gelatinous sheath; asexual reproduction by monospores.

Edwards (1969) cultured Texas isolates of this species and found them to reproduce asexually by monospores, which germinated into pluriseriate filaments.

These minute plants are common epiphytes on coarser algae during the warmer months.

ORDER BANGIALES

FAMILY ERYTHROPELTIDACEAE

Erythrocladia subintegra Rosenv. Fig. 2

Plants rose to red, microscopic epiphytes consisting of laterally compacted branched uniseriate filaments forming monostromatic discs; marginal cells often lobed or forked; asexual reproduction by monospores.

Nichols and Lissant (1967) and Edwards (1969) demonstrated in culture that this species reproduces asexually by monospores which germinate to produce monostromatic discs.

These minute plants are common epiphytes on coarser algae throughout the year.

Erythrotrichia carnea (Dillw.) J. Ag. Figs. 6-8

Plants red, microscopic, epiphytic, solitary, with unbranched filaments, to 0.5 cm long; initially uniseriate, but becoming biseriate in older parts, 20-30 µm diameter; cells subquadrate with a single stellate plastid; asexual reproduction by monospores cut off by an oblique wall near the distal end of the cell.

This minute epiphyte is common on coarser algae during the warmer months.

Erythrotrichia ciliaris (Carm. ex Harv.) Thur. in Le Jol. Fig. 4

Plants rose to brown, gregarious epiphytes, 3-5 cm long; filaments unbranched, initially uniseriate, 15-25 µm diameter, but becoming pluriseriate, 40-60 µm diameter; cells subquadrate to elongate with a single stellate plastid; asexual reproduction by monospores.

This species is found epiphytic on *Gelidium* and other eulittoral species during late winter and spring on rocks at

Beaufort, Topsail Inlet, and Masonboro Inlet. Its distinctive color and development from gregarious basal mats, as illustrated by Ardré (1970a), easily distinguish this species from the purple black, epilithic *Bangia atropurpurea* which grows above it.

According to Ardré (1970a), this species sometimes forms strap-shaped blades of up to 8-9 parallel rows of cells. Taylor (1960) similarly illustrates and describes *Erythrotrichia vexillaris* (Mont.) Hamel, which he reports from the American tropics to North Carolina. Williams (1948a) lists both *E. ciliaris* and *E. vexillaris* for Cape Lookout, but gives no indication of how the two entities can be distinguished. Since the descriptions for these two taxa (Taylor 1957, 1960) are readily accommodated by the expanded circumscription of *E. ciliaris* by Ardré (1970a), the suspicion must remain that in the New World they are regional epithets for the same entity. Ardré (1970a) suggests that *E. ciliaris* like *E. carnea* (Taylor 1957, 1960) is a plant of cosmopolitan distribution.

FAMILY BANGIACEAE

Bangia atropurpurea (Roth) C. Ag. (= *Bangia fuscopurpurea* [Dillw.] Lyngb.) Fig. 10

Plants purple to black, to 5 cm long, forming epilithic mats in the supralittoral; filaments initially uniseriate but becoming pluriseriate with 6-8 radial divisions; filaments 80-100 μm diameter; cells 2-4 times broader than tall; asexual reproduction by monospores.

During winter and early spring, this alga forms supralittoral mats on rocks at Beaufort, Topsail Inlet, Masonboro Inlet, and Fort Fisher.

Elsewhere this species is reported to alternate seasonally with a filamentous conchocelis phase (Richardson and Dixon 1968; Edwards 1969; Dixon 1970), but the fate of monospores in local material is unknown.

Geesink (1973) has shown in culture that *B. fuscopurpurea*, which is widely reported from the Americas, is conspecific with *B. atropurpurea*, previously assumed to be a freshwater species.

Porphyra carolinensis Coll et Cox Figs. 9, 11

Plants tufted, to 4 cm long, ovate-lanceolate; blades monostromatic with microscopic marginal teeth on older specimens; from a surface view cells in younger parts in parallel rows, eventually forming rosettes of angular, square-polygonal cells, 8-12 µm diameter, with a single-lobed plastid; sporulation from microscopic marginal patches; α-spores in packets of 16, β-spores in packets of 32; marginal monospores present.

This species grows in the supralittoral and upper eulittoral on rocks and coarser algae, achieving maximum development during spring and early summer, but persisting in reduced form throughout the year.

This entity has been various reported as *P. umbilicalis* (L.) Kuetz., *F. laciniata* (Lightf.) J. Ag., and *P. vulgaris* C. Ag. (Williams 1948a). The former species is now held to be synonymous with *P. purpurea* (Roth) C. Ag., while *P. vulgaris*, previously recorded from Florida (Taylor 1928), has

Systematic Account : 37

been reduced to synonymy with *P. linearis* Greville (Ardré 1970a). Although local collections seem to agree in most respects with the circumscription of *P. linearis* in the warm temperate waters of Portugal and Morocco (Ardré 1970a), Coll and Cox (1977), citing the distinctive marginal teeth and monospores of our plant as sufficient reason for distinguishing it from the above species, erected a new taxon, *P. carolinensis*. It would be of interest to determine the relationship of our dentate *Porphyra* to those reported from Japan (Kurogi 1972).

Porphyra rosengurtii Coll et Cox   Figs. 5, 12

Plants brown to purple with one-to-few blades, oblong-lanceolate when young, but becoming wide-spreading and lobed from a rounded base, to 30 cm long; blades monostromatic, 50-70 µm thick; cells in surface view oval to rectangular, 10-15 x 15-25 µm with a single-lobed plastid; α-spores in packets of 8; colorless β-spores in packets of 32-64; sporulation from deliquescing elongated patches parallel to the blade margin; monospores present.

This species occurs during winter and early spring in the littoral on jetties at Beaufort and Masonboro Inlet, as well as in the upper sublittoral on rocks and oyster reefs in sounds and estuaries.

Historically, large *Porphyra* specimens from North Carolina have been ascribed to either *P. umbilicalis* or *P. leucosticta* (Hoyt 1920; Williams 1948a, 1949; Taylor 1957, 1960), with the former receiving broadly lobed forms while ribbon-like plants were referred to the latter. Coll and Cox (1977) examined material from North Carolina and concluded that the specimens available to them could not be included in either of these taxa, although authentic P. umbilicalis is known to occur on the northern Atlantic coast of America.* Consequently, they erected a new taxon, P. rosengurtii to accommodate the North Carolina broadbladed entity.

It is generally assumed that the life histories of *Porphyra* species involve a filamentous conchocelis phase (Dixon 1973; Conway et al. 1975). To date no culture work has been published on North Carolina species, so their mode of reproduction remains unknown. Local plants sporulate in response to spring tides during the winter months, but the fate of the resultant α- and β-spores is unknown, as is the role of an hypothesized filamentous phase in their life history.

SUBCLASS FLORIDEOPHYCEAE

ORDER NEMALIALES

FAMILY ACROCHAETIACEAE

Members of the Achrochaetiaceae are simple or branched uniseriate filaments usually less than 5 mm tall; reproducing asexually by monospores, bispores, or tetraspores, or sexually with carpospores forming on outgrowths of the fertilized carpogonium which is devoid of a pericarp (Woelkerling 1973a). Taylor (1957, 1960), using a taxonomic scheme

---

*J. Coll 1979; personal communication.

based primarily on chromoplast morphology, lists five genera in this family from the North American Atlantic coast: *Rhodochorton, Kylinia, Audouinella, Colaconema,* and *Acrochaetium*. Woelkerling (1971, 1972b, 1973a, 1973b), however, has rejected the classification of genera based on chromoplast morphology because his studies showed plastid shape to be so variable in certain taxa that particular species could be placed in several genera simultaneously.

Several investigators have demonstrated considerable variation of presumed reliable taxonomic characters such as basal system morphology, branching morphology, and sporangial arrangement, all of which are used in the key for the local species (West 1968; Stegenga and Borsje 1976; Stegenga and Vroman 1976). Woelkerling (1971, 1972b, 1973a, 1973b) has suggested that most members of this family be included in the genus *Audouinella* if they possess evidence of sexual reproduction or in the form genus *Colaconema* in the absence of sexual reproduction. Although these taxonomic conclusions are not universally accepted (D'Lacoste and Ganesan 1972), they are followed in this study because the detailed illustrations and descriptions of Woelkerling (1972b, 1973a, 1973b) for adjacent regions provide the only critical means for determining North Carolina species. Presumed synonyms are given in the systematic account so that reference can be made to corresponding descriptions in other sources (Børgesen 1913-20; Hoyt 1920; Taylor 1928, 1957, 1960; Edwards 1970).

It should be mentioned that the taxonomy of audouinelloid algae is further confused by reports of acrochaetium-like plants alternating with certain macroscopic red algae, e.g., *Nemalion pulvinatum* (Umezaki 1967), *Scinaia complanata* (van den Hoek and Cortel-Breeman 1970), *Liagora farinosa* (von Stosch 1965), and *Pseudogloiophloea confusa* (Ramus 1969). Presumably these acrochaetium-like phases, where known, are encompassed by the form genus *Colaconema* sensu Woelkerling.

Acrochaetiaceae are a common and persistent aspect of the epiphytic flora of coastal North Carolina. These small plants are found on virtually all larger algae and seagrasses throughout the year. Too few specimens have been examined to generalize about seasonal periodicity or local distribution.

Audouinella alariae (Jonsson) Woelk. (= *Kylinia alariae* [Jonsson] Kylin) Figs. 15-17

Plants to 1 mm tall, arising from a swollen globose cell; branching secund to alternate; reproduction by 1-2 ovoid monosporangia on stalks; sexual reproductive structures not seen.

Audouinella dasyae (Coll.) Woelk. (= *Acrochaetium dasyae* Coll.) Figs. 13-14

Plants to 1 mm tall, arising from a persisting basal cell obscured by short rhizoidal filaments; cells 3-5 times longer than broad, containing a single parietal plastid and pyrenoid; reproduction by sessile, ovoid monospores borne in unilateral series in the upper branches; sexual reproductive structures not seen.

Audouinella hallandica (Kylin) Woelk. (= *Acrochaetium sargassi* Børg.) Figs. 18-20, 26

Plants to 1 mm tall, arising from the original spore persisting as a cylindrical unicellular basal cell; mature filament cells 2-3 diameters long, containing a single parietal plastid and pyrenoid; branching sparse; reproduction by ovoid monospores borne in series, 1-2 on lateral stalks throughout the plant; sexual reproductive structures not seen.

Audouinella microscopica (Naeg. in Kuetz.) Woelk. (= *Kylinia crassipes* [Børg.] Kylin) Figs. 21-22

Plants minute, to 100 μm tall; branching simple, arcuate; reproduction by sessile monosporangia borne in unilateral series; female plants with terminal carpogonia tipped with elongate trichogynes.

Audouinella secundata (Lyngb.) Dix. (= *Colaconema secundata* [Lyngb.] Woelk.) Figs. 23-25

Pseudoparenchymatous basal system giving rise to erect, branched filaments, to 2 mm tall; mature cells 3-5 times longer than broad, containing a single axial chloroplast and central pyrenoid; reproduction by ovoid monospores borne 2-3 on lateral stalks; sexual reproductive structures not seen.

Audouinella thuretii (Bornet) Woelk. (= *Acrochaetium thuretii* [Bornet] Coll. and Herv.) Fig. 27

Plants to 1 mm tall, arising from an extensive prostrate system of branching filaments which eventually coalesce into a disc; erect filaments irregularly branched, often ending in an elongate tapering cell; cells 2-3 diameters long, containing a lobed plastid and 1 pyrenoid; monosporangia solitary, sessile, or stalked.

FAMILY CHAETANGIACEAE

Scinaia complanata (Coll.) Cott.   Figs. 28,50

Plants pink to rose, firm to gelatinous, to 10 cm tall, repeatedly dichotomously branched; the branches cylindrical and blunt-tipped, nearly hollow at maturity; structurally derived from multiaxial medullary filaments which branch dichotomously in the assimilatory zone, eventually ending in large polyhedral surface cells.

A few sterile specimens have been collected from sublittoral rocks at Masonboro Inlet and adjacent near-shore ledges in late spring and summer. Van den Hoek and Cortel-Breeman (1970) cultured isolates of this species from the Atlantic coast of France and found the macroscopic gametophyte phase to alternate with a filamentous acrochaetium-like tetrasporophytic phase. They suggested that in the western Atlantic this species is "tropical to subtropical" (= eurythermal tropical) with the filamentous form acting as an overwintering phase in its northern range.

ORDER GELIDIALES

FAMILY GELIDIACEAE

Members of the Gelidiaceae are stiff to cartilaginous

Systematic Account : 43

plants with wiry axes, internally differentiated into a
large-celled medulla and a peripheral small-celled cortex
with thick-walled filamentous rhizines more or less abundant
in the medulla and subcortex (Dixon 1958). Cruciately
divided tetrasporangia are borne in the cortex of branchlets.
The two most common genera, *Pterocladia* and *Gelidium*, are
distinguished by the number of locules and ostioles in the
carposporophyte (Taylor 1957, 1960; Edwards 1970).

After working with British *Gelidium* populations, Dixon
(1961, 1966) characterized the genus as one of the most confused of the red algae, nomenclaturally and taxonomically.
Stewart (1968) found a similar situation with Pacific species
of *Pterocladia* in southern California where seven species
were found to be conspecific with *Pterocladia pyramidale*,
which in turn was determined to be conspecific with European
P. capillacea. More recently, Santelices (1976) examined
material of *Pterocladia americana* and found it to belong to
the genus *Gelidium* and proposed a new combination: *Gelidium
americanum* (Tayl.) Santilices.

In coastal North Carolina, five species of *Gelidium*
have been recorded (Hoyt 1920; Williams 1948a; Wiseman 1966;
Taylor 1960) in addition to *Gelidium americanum* (= *Pterocladia americana*). It is probable that most of these reports
concern growth forms of two species. For example, Taylor
(1960) lists *Gelidium corneum* (Hudson) Lamour. as a doubtful
species for warm temperate and tropical American waters and
considers all but two records untrustworthy. Likewise, the
reports of *Gelidium pulchellum*, a Mediterranean species, in
North Carolina (Williams 1948a; Wiseman 1966) must be treated

with suspicion. Although Williams gives several morphological criteria for distinguishing between *G. pulchellum* and *G. crinale*, in the light of the extreme morphological variation demonstrated in the Gelidiaceae (Stewart 1968), it seems probable that North Carolina material can easily be accommodated by Taylor's circumscription of *G. crinale* (1960) for our coast. Finally, both Santelices (1976) and Taylor (1960) consider *Gelidium caerulescens* Kützing (= *Pterocladia caerulescens* [Kuetz.] Santelices to be a Pacific species (type from New Caledonia) and suggest that most American records are referable to *Gelidium americanum*.

Gelidium crinale (Turn.) Lamour. Figs. 30-33, 36

Plants reddish purple to brown, forming dense turf on wave-swept rocks; individuals terete and irregularly branched or slightly flattened and pinnately branched; axes slender, less than 0.5 mm diameter, to 3-5 cm long in more sheltered locations and to 15 cm long in exposed sites; in transection showing a large-celled medulla transversed by thick-walled rhizines; in longitudinal section showing medullarly filaments closely packed and parallel; tetrasporangia borne in terminal branches or lateral spatulate branches; bilocular cystocarps solitary, embedded in spindle-shaped swollen branchlets.

This alga is one of the most common in our area, achieving maximum development in the eulittoral during winter and spring, but persisting throughout the year. Attention has been called above to the extreme morphological plasticity exhibited by this species including the slender, elongate, slightly branched form common on Masonboro Inlet jetty; the

more robust, shorter, highly branched turf form on Fort
Fisher rocks and at Beaufort; and the minute pulvinate form
which develops in the Cape Fear estuary and Federal Basin at
Fort Fisher.

Gelidium americanum (Tayl.) Santelices (= *Pterocladia americana* Tayl.) Figs. 29, 34-35

Plants reddish purple, 1-3 cm tall; main axes terete or flattened, with short, pinnate, spatulate branchlets always flattened; in longitudinal section showing a few parallel filaments giving rise to short branching filaments in the cortical area, and ending in an epidermal layer at the surface; tetrasporangia borne in lateral spatulate swollen branchlets.

This species is much less abundant than *G. crinale*, occurring in small patches in the upper sublittoral at Radio Island jetty at Beaufort, and Federal Basin at Fort Fisher where it achieves best development in summer and fall. Its restricted distribution in the study area seems to indicate a tolerance for protected, hyposaline, and turbid environments.

ORDER CRYPTONEMIALES

FAMILY SQUAMARIACEAE

Peyssonnelia rubra (Grev.) J. Ag. Figs. 40-43

Plants light to dark red, membranous to calcified; firmly adhering to substratum by numerous short rhizoids;

hypothallus distromatic in transverse section, consisting of dichotomous rows of cells; perithallus of ascending simple or once-branched filaments; cystocarps scattered, consisting of a few large carpospores; tetrasporangia tetrapartite, borne laterally on erect filaments.

This species is abundant during summer and fall in the upper sublittoral on rocks at the Radio Island and Masonboro Inlet jetties and has been collected offshore to depths of 60 m (Schneider 1976).

Although two additional species of *Peyssonnelia* have been reported for coastal North and South Carolina, *P. polymorpha* (Williams 1948a; Wiseman 1966) and *P. conchicola* (Williams 1949), all of the specimens observed in this study could be included in *P. rubra*.

FAMILY CORALLINACEAE

The Corallinaceae have a multiaxial type of construction with branching filaments forming pseudoparenchymatous crusts or erect axes. Reproductive structures are borne in distinct superficial conceptacles. Tetrasporangia are zonate. Plants in this family deposit calcium carbonate (calcite) in their cell walls, giving them a pinkish red or even whitish color. Historically, these algae have been divided into two groups: crustose (nonarticulated) forms and erect (articulated) forms. Decalcification is advisable and, indeed, essential for the crustose forms before identification can be attempted. Review papers about both crustose forms (Littler 1972) and the articulated forms (Johansen 1974) are available, as is an analysis of taxonomically significant morphological features for both groups (Johansen 1976).

Dermatolithon pustulatum (Lamour.) Fosl. (= *Lithophyllum pustulatum* [Lamour.] Fosl.)   Fig. 39

Plants minute, epiphytic crusts, structurally of 1-3 layers of vertically elongate cells which cut off lenticular cover cells on the crust surface; zonate tetrasporangia formed in conspicuous conceptacles.

This species is common during the warmer months; it is epiphytic on coarser algae and on *Zostera* (Brauner 1975) in shallow bays.

Heteroderma lejolisii (Rosanoff) Fosl. (= *Fosliella lejolisii* [Rosanoff] Howe)   Figs. 37-38

Plants minute, epiphytic, monostromatic crusts; conceptacles subhemispherical, crowded or nearly confluent.

This epiphytic alga is found mixed with *Dermatolithon pustulatum* on coarser algae and *Zostera*. *D. pustulatum* typically forms confluent crusts while *H. lejolisii* forms crusts which are never confluent but readily overgrow one another. Although these distinctions are obscured in densely crowded specimens, transections of decalcified plants permit ready identification of the respective species.

Amphiroa beauvoisii Lamour.   Fig. 46

Plants calcified and articulated, forming rose red tufts to 5 cm tall; branching dichotomous; segments at the bottom are cylindrical, 600 µm diameter or more, segments at the top are compressed and bifurcate at the tips; conceptacles scattered over the surface of the segments. This alga is occasionally encountered sublittorally on coastal jetties or on rocks in shallow coastal water, but is apparently much more common offshore (Schneider 1976).

Corallina cubensis (Mont.) Kuetz. Fig. 45

Plants calcified and articulated, to 3 cm tall; compressed main axes beset with irregularly to pinnately branched cylindrical segments.

This alga is occasionally encountered in near-shore coastal waters on rock ledges, growing with *Sargassum* and *Dictyopteris*. Like *Amphiroa beauvoisii*, it is more abundant in offshore waters at moderate depth (Schneider 1976).

Jania adhaerens Lamour. Fig. 44

Plants calcified and articulated, less than 2 cm tall; axes terete, to 100 μm diameter, branching regularly dichotomous and wide-angled; segments to 4 diameters long.

This alga has been collected on basal crusts of *Sargassum* and *Dictyopteris* in near-shore waters but is more abundant in offshore waters of moderate depth (Schneider 1976).

FAMILY GRATELOUPIACEAE

Grateloupia filicina (Wulf.) C. Ag. Figs. 48, 53

Plants dark red, to 5 cm tall; main axis simple or irregularly branched, compressed and beset with numerous pinnate branches; structurally of anastomosing filaments which become compacted, forming a cortex of anticlinal cell rows.

This alga is uncommon on coastal jetties, but is often found during summer in the upper infralittoral at Cape Lookout jetty. In North Carolina this alga bears little resemblance to plants encountered in the American tropics (Taylor 1960) or the Gulf of Mexico (Edwards 1970). In local waters *Grateloupia* is small, rubbery, and *Pterocladia*-like rather than large and gelatinous, as found elsewhere. Transections are thus essential to confirm the identity of any material in question.

Halymenia floridana J. Ag.   Figs. 49, 51

Plants pinkish red, 4-10 cm tall, at first ovate but becoming palmate with age; blades membranous, 100-150 μm thick; medulla of irregularly branched filaments of various sizes, forming numerous large, stellate ganglia; cortex 1-3 cells thick, the cells subquadrate in transection.

Halymenia gelinaria Coll. and Howe   Figs. 47, 52

Plants red to purple, to 25 cm tall, obovate, very gelatinous, 300-500 μm thick; medulla of pillarlike conspicuously segmented vertical filaments; subcortex of small-celled filaments, frequently anastomosing to form ganglia.

Searles and Schneider (1978) have suggested that the *H. gelinaria-H. floridana* complex in North Carolina constitutes a single species. Both of these taxa and the four additional *Halymenia* species reported for the state (Searles and Schneider 1978) are in need of critical taxonomic reevaluation.

ORDER GIGARTINALES

FAMILY GRACILARIACEAE

Members of the Gracilariaceae are large, cartilaginous plants with terete or flattened, highly branched axes, internally differentiated into a medulla of large, colorless cells surrounded by a cortex of much smaller, pigmented cells. Tetrasporangia are cruciately divided and scattered in the subcortex. Spermatia are formed from cortical cells in patches. Cystocarps are hemispherical and protuberant, opening by a prominent ostiole.

Six species of *Gracilaria* have been recorded for North Carolina (Taylor 1960; Schneider 1975c, 1976), but only four of these are commonly found in inshore waters.

Gracilaria blodgetti Harv. Figs. 54, 61, 67

Plants rose red, to 20 cm tall; branching radial to several orders, main axes terete, 2-3 mm diameter, ultimate branchlets characteristically constricted at the base; medulla of large, subspherical thin-walled cells surrounded by a cortex of smaller cells in anticlinal rows; tetrasporangia restricted to the branchlets.

This plant is occasionally collected in the drift or on near-shore rock ledges, but is primarily a species of deeper water (Schneider 1976).

Gracilaria foliifera (Forssk.) Børg. Figs. 55, 58

Plants to 20 cm tall, dark red to olive green, flattened above into straplike blades to 1.5 cm wide; branch tips

Systematic Account : 51

acute; branching in the plane of the blade; medulla of large, thick-walled, oblong cells surrounded by a cortex of 1-2 layers of smaller cells; tetrasporangia scattered in the cortex; cystocarps protuberant over the entire surface of the plant.

This species achieves maximum growth and reproductive development during the warmer months, but is a common member of the flora throughout the year in both coastal and estuarine habitats.

Recent studies of *Gracilaria* populations in New England and maritime Canada have shown that specimens there previously referred to as *G. foliifera* and *G. verrucosa* constitute a single polymorphic species (Edelstein 1977; Chapman et al. 1977; Edelstein et al. 1978) which has been given the name *G. tikvahiae* (McLachlan 1979). Although authentic specimens of *G. verrucosa* (McLachlan 1979) and *G. foliifera* (Chapman et al. 1977) have been identified from England, the occurrence of these taxa in the northwestern Atlantic is considered doubtful (McLachlan 1979). The familiar epithets *G. foliifera* and *G. verrucosa* have been retained in this study, however, until the status of North Carolina populations can be determined.

Gracilaria sjoestedtii (Kylin) Papenf. (= *Gracilariopsis sjoestedtii* [Kylin] Dawson)   Figs. 59, 60, 70

Plants dark red, to 50 cm long or more; lax, terete; axes emergent from sublittoral sand, 2-3 mm diameter, short branchlets arise radially, irregularly from distinct main axes; medulla of large, thin-walled cells surrounded by a cortex of up to 4-5 cells in anticlinal rows; large cystocarps strongly protuberant.

This species achieves maximum vegetative and reproductive development during the late spring and summer, but persists throughout the year in the upper sublittoral coquina and sand at the Fort Fisher beach.

Gracilaria verrucosa (Huds.) Papenf. Figs. 56-57

Plants purple to reddish brown, to 20 cm tall; branching radial to several orders, main axes terete, 2-3 mm diameter, tapering to the ultimate branches which form slender tips; medulla of large, oblong, thin-walled cells surrounded by a cortex of 1-2 layers of small cells.

This plant is common during the warmer months in bays and estuaries, where it grows on oyster shells and in unattached, loosely entangled tufts.

FAMILY SOLIERIACEAE

Members of the Solieriaceae are large, fleshy to cartilaginous plants with cylindrical or compressed axes, or broadly lobed blades, structurally comprised of a clearly filamentous central medulla, often merging into a peripheral large-celled zone below the cortex of anticlinally branched cell rows. Tetrasporangia zonately divided and scattered in the subcortex; cystocarps buried in the central medulla, strongly projecting or in papillae.

In North Carolina, this family is represented by four genera: *Euchema*, *Solieria*, *Meristotheca*, and *Neoagardhiella*.

Only the latter two were found in coastal waters during this study.

Neoagardhiella baileyi (Harv. ex Kuetz.) Wynne and
Taylor (= *Agardhiella baileyi* [Harv. ex
Kuetz.] Tayl.) Figs. 62-64, 71-74

Plants rose red to straw colored, to 30 cm long; terete throughout, main axes radially branched to several orders, branches basally constricted and attenuated towards the tips; medulla of large, subspherical cells surrounding a filamentous cavity; cortex compact, with inner cells grading to an outer layer of 1-2 cells; cystocarps swollen, embedded in main axes and larger branches; tetrasporangia scattered in the subcortex; spermatia formed in patches from elongated cortical cells.

This species exhibits a wide range of morphological variation. Some specimens are cartilaginous, dark red, beset with blunt branchlets, and bearing cystocarps in papillae as in species of *Eucheuma* (Fig. 62). Other plants are less rigid, but not as soft as is typical for the species *Neoagardhiella baileyi*, and tend to branch in one plane rather than radially, with compressed branches (Fig. 64).

Neoagardhiella baileyi achieves maximum development during winter and late spring, but persists in reduced form throughout the year.

Meristotheca floridana Kylin Figs. 65, 68

Plants rose red, palmately or irregularly lobed, fleshy to gelatinous, to 25 cm long; blades 300-400 μm thick, structurally with a loose filamentous medulla, connecting with large, irregularly shaped cells of the inner cortex; outer cortex of two layers of small cells; cystocarps swollen in papillae scattered over the blade surface.

This species is encountered in the drift after spring storms, apparently torn loose from offshore outcroppings where it occurs in abundance (Schneider 1976). Occasional attached specimens have been collected in the sublittoral on the Masonboro Inlet jetty.

Vegetative *Meristotheca* plants closely resemble *Halymenia gelinaria* in color, texture and habit. However, they are structurally distinct and can be readily distinguished in transection.

FAMILY HYPNEACEAE

Members of the Hypneaceae are bushy, irregularly branched, stiff to cartilaginous plants, structurally showing a persistent central filament which can be seen in the center of the large-celled medulla. Apical growth is from a single recognizable and persistent cell terminating each branch. Tetrasporangia are zonate, borne in swollen bands on the base of branchlets in the cortical layer of anticlinal cell rows. Cystocarps are swollen and sessile on the main axes.

Four species of *Hypnea* are reported in North Carolina: *H. cervicornis*, *H. volubilis*, *H. cornuta*, and *H. musciformis* (Taylor 1960; Schneider and Searles 1976; Schneider 1976), but only the latter two have been collected inshore.

Hypnea cornuta (Lamour.) J. Ag.   Figs. 80-82

Plants with one-to-few main axes, to 10 cm tall; alternately branched, more or less covered with pedicellate and stellate branchlets; crozier tips lacking; sessile cystocarps urn-shaped.

This species is much less common than *H. musciformis*, but occasional specimens were collected during spring and summer in the bays behind Wrightsville Beach and on the Masonboro Inlet jetty.   It is also reported to be common during summer at Fort Macon.*

Hypnea musciformis (Wulf.) Lamour.   Fig. 75-79, 83

Plants to 20 cm long, bushy, often entangled, densely branched, alternately and radially; some branches with thickened, curved (crozier) tips; branchlets simple, elongate and tapering to the tip; cystocarps sessile, spherical.

This species is common during the warmer months on floating docks in the sounds, and in the eulittoral on jetty rocks where it forms extensive mats.

FAMILY PHYLLOPHORACEAE

Gymnogongrus griffithsiae (Turn.) Mart.   Figs. 69, 84

Plants dark red to brown, bushy, repeatedly forking, to 5 cm tall, forming pulvinate tufts; medulla of angular, thick-walled, densely placed cells; cortex of anticlinal cell rows; tetrasporangia cruciate, borne in series on short

---

*C. W. Schneider 1970:   personal communication.

filaments in swollen nemathecia.

This species is especially abundant in Federal Basin where it forms unattached tufts in the shallow water and at Fort Fisher on eulittoral rocks. This perennial becomes reproductive during the late spring and summer.

FAMILY GIGARTINACEAE

Gigartina acicularis (Wulf.) Lamour. Figs. 66, 85

Plants olive to black, entangled, cartilaginous, to 8 cm tall; main axes repent, branching alternate to secund near the apices; medulla pseudoparenchymatous, of round, thick-walled, loosely placed cells, grading into inner cortex of irregularly shaped cells; outer cortex of anticlinal cell rows.

This species is occasionally collected unattached in the same habitats as *Gracilaria verrucosa* in the bays behind Wrightsville Beach. It also has been collected attached in the upper infralittoral at Cape Lookout jetty.*

ORDER RHODYMENIALES

FAMILY RHODYMENIACEAE

Rhodymenia pseudopalmata (Lamour.) Silva Figs. 86, 98-99

Plants rose to red, to 12 cm tall with narrow membranous dichotomously branching blades; in transection showing a

---

*C. W. Schneider 1979: personal communication.

large-celled medulla surrounded by a small-celled pigmented cortex; prominent cystocarps typically restricted to blade margins; cruciately divided tetrasporangia in yellowish sori behind the frond tips.

This species is a perennial in the sublittoral on the Radio Island and Masonboro Inlet jetties, where it achieves maximum growth and reproduction during spring and summer.

Rhodymenia pseudopalmata and *R. divaricata* (Schneider and Searles 1976) are the only species in the genus *Rhodymenia* occurring on the American east coast because *Rhodymenia palmata* (= *Palmaria palmata* [L.] O. Kuntze) has been removed to another genus (Guiry 1974a, 1974b). Dawson (1941) provides detailed illustrations of vegetative and reproductive structures in his monographic treatment of *Rhodymenia* in the Gulf of California.

Botryocladia occidentalis (Børg.) Kylin   Fig 87

Plants to 20 cm long, with wiry axes more or less abundantly branched, bearing rose red, shortly pedicellate, subspherical bladders 4-8 mm long; reproductive structures not seen.

In North Carolina this species is common in the drift and is found attached on near-shore rock ledges at depths of 5-10 m. *Botryocladia pyriformis* has been reported from offshore waters (Schneider 1976), but was not encountered in this study.

FAMILY CHAMPIACEAE

Champia parvula (C. Ag.) Harv.   Figs. 90-92

Plants rose to yellow, soft and tufted to 10 cm long;

branching alternate and dense; branches visibly consisting
of barrel-shaped hollow segments partitioned by septa;
tetrahedral tetrasporangia scattered over the surface of
branches; cystocarps in prominent flask-shaped sessile
pericarps.

This species is perennial in North Carolina and is
especially abundant on sublittoral rocks at Radio Island
and Masonboro Inlet jetties, where it achieves maximum
vegetative and reproductive development during spring and
early summer in shaded and protected sites. *C. parvula* var.
prostrata L. Williams is the common form of this species
offshore (Williams 1951; Schneider 1976).

Lomentaria baileyana (Harv.) Farl. Figs. 88, 93-94

Plants rose to red, soft, to 6 cm long, densely branching and creeping, forming tufts or mats; cylindrical branches irregularly secund; axes hollow except at branch bases which are septate; tetrahedral tetrasporangia scattered over the surface of branches; cystocarps in prominent sessile, clustered pericarps.

This species is perennial in North Carolina on sublittoral jetty rocks where it forms mats at the low water mark in shaded and protected sites. Maximum vegetative and reproductive development occur in spring and early summer.

Lomentaria orcadensis (Harv.) Coll., also reported for
North Carolina (Williams 1948a and Hoyt 1920 as *L. rosea*;
Schneider 1976), was not encountered in this study.

ORDER CERAMIALES

FAMILY CERAMIACEAE

Members of the Ceramiaceae consist of branched, uniseriate filaments, often with nodal cortication which can spread over the internodes, resulting in a parenchymatous appearance. Tetrasporangia cruciately or tetrahedrally divided; sessile or on unicellular stalks in uncorticated species or located in the cortex of corticated species. Cystocarps borne superficially and naked except for subtending involucral branchlets. Antheridia consisting of clusters of determinate branchlets or forming patches on the nodes of corticated species.

Anotrichium barbatum (C. Ag.) Naeg. (= *Griffithsia barbata* C. Ag.) Figs. 101-102

Plants rose red, 2-3 cm tall; erect filaments sparingly, subdichotomously branched; cells swollen, 5 or more diameters long; younger cells supporting conspicuous whorls of thrice-branched trichoblasts; tetrasporangia borne several to a node on branchlet cells; antheridia conical, borne 1-2 at fertile nodes on elongate branchlet cells.

This species, epiphytic on pelagic *Sargassum*, is infrequently collected at Wrightsville Beach during the summer months. It is previously unreported for North Carolina and probably fails to perenniate on our coast.

Anotrichium tenue (C. Ag.) Naeg. (= *Griffithsia tenuis* C. Ag.) Figs. 100, 104

Plants rose red, forming tufts to 3 cm tall from a

system of creeping filaments attached by unicellular rhizoids; erect filaments simple or sparingly alternately branched; cells rectangular, swollen, 100 x 400 μm, slightly constricted at the nodes; apical cells caplike.

This species is infrequently collected during the summer months at Masonboro Inlet jetty in the lower eulittoral, and is previously reported (as *Griffithsia tenuis*) from Radio Island (Blomquist and Humm 1946) and offshore (Schneider 1976).

Baldock (1976), in a monograph of the genus *Griffithsia*, separated *G. tenuis* and *G. barbata* from this taxon and referred them to *Anotrichium* on the basis of several distinctive reproductive characteristics.

Antithamnion cruciatum (C. Ag.) Naeg. Figs. 105-107

Plants rose red, 1-2 cm tall; uncorticated uniseriate axes erect from creeping stoloniferous filaments; lateral branches with a distinct spherical basal cell, arising in opposite pairs from each axial cell; branch tips brushlike with nearly secund branchlets; tetrasporangia cruciately divided, adaxial on short one-celled pedicels; antheridial branches of 4-8 cells, globular, terminal on main axes.

This species grows in association with *Callithamnion byssoides* and *Ceramium fastigiatum* at the low tide mark on jetty and breakwater rocks at Radio Island and Masonboro Inlet. Maximum vegetative and reproductive development occur during the warmer months. Recent culture studies of this alga (Kapraun 1977b) indicate that temperature is of primary importance in its seasonality in North Carolina. Whittick and Hooper (1977) demonstrated that the prostrate variety *radicans*, previously reported from North Carolina (Blomquist and Humm 1946; Schneider 1976), can develop into the typical erect form, thus eliminating the need for retention of varietal status for these growth forms.

Callithamnion byssoides Arn. ex. Harv. in Hook. Figs. 108-110

Plants light pink, soft, plumose, to 10 cm tall; uncorticated, uniseriate axes pinnately, alternately branched, branchlets upcurved and rounded toward the tip; sessile tetrasporangia tetrahedrally divided, obovate, 30 x 40 μm; spermatia borne on short, compact filaments in adaxial tufts; paired carposporophytes subspherical, borne laterally on the main axis.

This species grows in the lower eulittoral on jetty rocks where it achieves maximum vegetative development in spring, becoming sexually reproductive during early summer. Recent culture studies on this alga (Kapraun 1978a) indicate that temperature and light intensity are primary causal factors in its periodicity in North Carolina.

Ceramium byssoideum Harv. Figs. 111-116

Plants light red, minute epiphytes on coarser algae; erect axes form an extensive creeping system of decumbent filaments; apices straight, nodes of 4-6 rows with the lowermost of transversely elongated cells; tetrasporangia 1-2 per node, strongly emergent; cystocarps spherical, borne laterally on older nodes, subtended by several elongate involucral branches.

This species is found epiphytic on *Codium* and *Padina*

during the summer. It is probably more common than published
reports for our area would suggest, but its small size and
association with numerous other minute red epiphytes make it
rather inconspicuous.

This entity has been reported from North Carolina as *C.
byssoideum* (Taylor 1960) and *C. transversale* (Williams 1951).
Feldmann-Mazoyer (1941) considered both of these taxa synonymous with *C. gracillimum* var. *byssoideum* (Harv.) G. Mayoz.
Recently, Womersley (1977) proposed that all of these taxa
be reduced to synonymy with *C. flaccidum* (Kuetz.) Ardissone.

Ceramium diaphanum (Roth) Harv. Figs. 117-124

Plants bushy, lax, to 20 cm tall, dull brownish red;
branching dichotomous, often with numerous adventitious
branchlets; branch tips forcipate; nodes with 4 or more rows
of irregularly placed cells of various sizes, often partially
covering deeply placed larger cells; tetrasporangia immersed
in the middle or upper part of the nodes; cystocarps spherical,
borne laterally on older nodes, subtended by one-to-few short
involucral branchlets; spermatia formed in cortical patches
on younger nodes.

This is the most common littoral *Ceramium* species in
North Carolina, occurring in a wide range of coastal and estuarine habitats. It achieves maximum vegetative and reproductive development during late winter, but persists in
reduced form throughout the year.

Ceramium strictum is considered distinct from *C. diaphanum* by Taylor (1957, 1960), and both taxa are reported
from North Carolina (Searles and Schneider 1978). In the

Systematic Account : 63

present paper, however, the European treatment is followed, which includes *C. strictum* as a varietal form of *C. diaphanum* (Feldmann-Mazoyer 1941; Gayral 1966; Ardré 1970a; Rueness 1978).

Ceramium fastigiatum Harv. in Hook. Figs. 126-130

Plants rose red, forming dense tufts, to 2 cm tall; erect axes from an extensive creeping system of horizontal filaments attached to the substratum by 1-to-2 celled rhizoids; apices straight or slightly incurved but not forcipate; nodes of 2 transverse rows with the lower usually larger than the upper; tetrasporangia one-to-few per node, strongly emergent; cystocarps spherical, borne laterally on older nodes, subtended by one-to-few involucral branches; spermatia formed in cortical patches on younger nodes.

This alga is common at lower water mark on jetty rocks at Radio Island and Masonboro Inlet during summer. In addition, Schneider (1976) reported it as abundant in offshore waters.

Ceramium rubrum (Huds.) C. Ag. Figs. 125, 131-135

Plants dark red, bushy, erect, to 25 cm tall; axes dichotomously branched; cortication complete except near the apices where axial internodal cells can be observed; apices forcipate; tetrasporangia immersed in nodal bands; cystocarps globular, on lateral branches subtended by upcurving involucral branchlets.

This species is perennial at Federal Basin near Kure Beach, where it grows in shallow, brackish water. It has also been dredged from the intercoastal waterway and is probably common in tidal creeks and estuaries throughout the study area. Maximum vegetative and reproductive development occur during March and April.

Comparative studies of *C. rubriforme*, *C. areschougii*, and *C. pedicellatum* from Nova Scotia have suggested that these entities in the northwestern Atlantic are probably synonymous with *C. rubrum* (Garbary et al. 1978).

Griffithsia globulifera Harv.   Figs. 95, 97, 103

Plants tufted, rose red, to 6 cm tall; consisting of branching uniseriate filaments of large, globular, multinucleate cells; axes uncorticated and bearing whorls of branching trichoblasts, especially in the branch tips; tetrasporangia tetrahedrally divided, borne in dense clusters at fertile nodes, surrounded by short, blunt, involucral cells.

This species has not been found attached when collected from inshore waters, but is occasionally encountered in the drift in southern Onslow Bay and is reported from moderate depth offshore (Schneider and Searles 1975; Schneider 1976).

Spermothamnion investiens (Crouan Frat.) Vick.   Fig. 96

Plants minute, forming a pink felt, to 6 mm tall; erect uniseriate filaments simple or sparingly alternately branched, arising from an extensive creeping system of filaments attached by unicellular rhizoids; cells rectangular, 25 x 100 µm; not constricted at the nodes; apical cells similar to other filament cells.

Hoyt (1920) reported this species very abundant on Bogue Beach near Beaufort where reproductive material was found throughout the year. In this study, the plant, epiphytic on coarser algae, was infrequently collected on subtidal jetty rocks at Masonboro Inlet.

Spyridia hypnoides (Bory) Papenf. (= *Spyridia aculeata* [Schimp.] Kuetz.)   Figs. 137-138

Plants light red, erect, bushy and broadly pyramidal, to 20 cm tall; branching radial, alternate from several main axes; branches completely corticated with alternating bands of narrow, elongate cells and shorter, broader cells; main branches often terminating in swollen, naked, hooked tips; branches setaceous with uniseriate ramelli having rings of nodal corticating cells; ramelli tipped with a terminal spine and none-to-few lateral unicellular spines.

This species was collected during summer on the jetties at Radio Island and Masonboro Inlet. It was occasionally encountered adrift or loosely attached to vegetation in the sounds where, in its lack of hooked tips and lax habit, it superficially resembled *S. filamentosa*.

Spyridia aculeata (Taylor 1960) is regarded as a synonym for *S. hypnoides* (Papenfuss 1968), but the relationship of *S. aculeata* var. *hypnoides* J. Ag., widely reported from the North Atlantic, to *S. hypnoides* is unknown.* In our area, *Spyridia hypnoides* and *S. filamentosa* can be readily distinguished on the basis of cortication, with the former having alternating bands of different-sized cells, as described above, and the latter having alternating bands of short and long cells of equal diameter. Identifications

---

*C. F. Papenfuss 1979:   personal communication.

based solely on the presence of terminal recurved spines, a
highly variable and thus unreliable taxonomic character, have
often led to misidentifications. Blomquist and Humm (1946)
examined material which Hoyt (1920) identified as *S. filamentosa* and found it to be *S. hypnoides* (as *S. aculeata*).
Although *S. filamentosa* is recorded for offshore waters
(Schneider 1976), inshore reports should be treated with
caution.

FAMILY DELESSERIACEAE

Members of the Delesseriaceae consist of membranous
blades, one-to-few cells thick. Apical cell divisions typically yield a median cell row, which in turn produces lateral
cell rows of determinant length, resulting in a blade appearing to have a midrib and veins. Tetrasporangia are tetrahedrally divided and borne in superficial sori. Cystocarps are
covered by a thin pericarp with a prominant ostiole. Spermatia are formed in patches.

Branchioglossum minutum C. Schn. Fig. 148

Plants rose red, minute, consisting of clusters of
blades to 2 cm long, arising from a discoidal base; each
blade with a distinct midrib and proliferous marginal branching; tetrasporangia formed in circular, opposite patches adjacent to the midrib in the apical end of the blade; cystocarps large, 300-500 μm diameter, spherical, protuberant,
forming on the midrib; spermatia formed in irregularly
shaped patches between the blade midrib and margin.

Systematic Account : 67

This species is not common in coastal waters, but every summer has been collected in small numbers growing with *Hypoglossum* under overhanging rocks at the Masonboro Inlet jetty, at low water mark or slightly above it.

This species was described by Schneider (1975b) as endemic to the North Carolina continental shelf and is herein newly reported for inshore waters. However, *Branchioglossum prostratum* (Schneider 1974), also described from dredged material and seemingly endemic to our offshore waters, was not encountered in the present study.

Caloglossa leprieurii (Mont.) J. Ag. Fig. 153

Plants purplish brown, forming creeping entangled mats; blades to 2 mm wide, repeated dichotomous forking from blade tips; growth from a distinct apical cell producing a distinct midrib in a monostromatic blade; reproductive structures not seen.

This species is common during winter on *Spartina* culms in the Cape Fear estuary near Kure Beach and has been collected from stumps in Pamlico Sound. *Caloglossa* is a brackish water plant, often growing on mud with *Bostrychia* and Ulvales where its dark color and small size make it easily overlooked.

A detailed study of the structure and reproduction of this species has been made by Papenfuss (1961).

Calonitophyllum medium (Hoyt) Aregood (= *Hymenena media* [Hoyt] Tayl.) Fig. 143

Plants rose, to 20 cm tall; erect blades membranous, repeatedly irregular, dichotomous branching, 1-2 cm wide;

margins wavy or proliferously dentate; monostromatic blades
with macroscopic veins in older parts and microscopic veinlets throughout; tetrasporangia borne in superficial sori
over the entire blade; cystocarps swollen, covered by a
pericarp, scattered over the blade surface; spermatia formed
proliferously in extensive patches.

This species is abundant in inshore waters in the upper
sublittoral at Masonboro Inlet where it is collected; it is
reproductive throughout the year. In addition, it is widely
reported from drift and dredged specimens (Hoyt 1920 as
Nitophyllum medium; Taylor 1960 and Schneider 1976 as *Hymenena
media*), indicating extensive deep water populations as well.

A recent morphological study of this species (Aregood
1975) has provided detailed information on the development
of its vegetative and reproductive structures as well as its
relationship to other Delesseriaceae.

Grinnellia americana (C. Ag.) Harv. Fig. 144

Plants pink to red, consisting of one-to-few erect,
lanceolate blades from a short stalk; blades to 20 cm long,
with a raised midrib through most of its length; tetrasporangia in elevated sori; cystocarps hemispherical with a
thin-walled pericarp.

This species is common during winter and spring in the
upper sublittoral at Lockwood's Folly Inlet, Masonboro Inlet,
and Radio Island. Although maximum vegetative and reproductive development occur during the colder months, isolated
individuals can be found throughout the year in sublittoral
habitats.

Hypoglossum tenuifolium (Harv.) J. Ag. var. *carolinianum*
L. Williams Figs. 145-147, 149-151, 154

Plants red, minute, gregarious, consisting of clusters of blades, forming entangled mats; blades to 3 cm long, oblong-lanceolate, branching from the conspicuous midrib, in pairs at regular intervals; tetrasporangia in oblong sori near the apex; cystocarps spherical, conspicuous on the midrib; spermatia formed in narrow patches toward the margins.

This species is particularly abundant at low water mark on rocks at Masonboro Inlet, where it typically forms spongy mats with *Champia parvula* and *Lomentaria baileyana*. Maximum vegetative and reproductive development occur during summer, but plants in reduced form persist throughout the year.

FAMILY DASYACEAE

Dasya baillouviana (Gmel.) Mont. Figs. 155, 141-142

Plants red to purple, bushy, freely branched, to 50 cm long; the main axes with 5 pericentral cells completely corticated by rhizoidal filaments, and supporting whorls of branchlets dividing into monosiphonous ramelli; tetrasporangia borne in elongate, stalked stichidia; stalked pericarps large, urn-shaped.

This species is abundant during winter months on rocks in the sublittoral at Radio Island, Masonboro Inlet, and Lockwood's Folly Inlet. Its large size and deep red color make this one of the most spectacular algae in our flora.

FAMILY RHODOMELACEAE

Members of the Rhodomelaceae consist of radially branched plants structurally derived from a persistent apical cell, which produces a distinctive central cell row. Pericentral cells cut off from the central filament result in a polysiphonous structure which may ultimately be superficially corticated by subsequent pericentral cell divisions or rhizoidal downgrowths. Trichoblasts (branched colorless hairs) are typically present in association with reproductive structures or actively growing apices. Tetrasporangia are formed beneath the pericentral cells in slightly modified filaments or in some cases in swollen branchlets or stichidia. Cystocarps are enclosed by an ostiolate pericarp and borne in branchlets. Spermatangia form in cones or platelets derived from modified trichoblast initials.

Bostrychia radicans (Mont.) Mont. (= *Bostrychia rivularis* Harv.) Fig. 226

Plants purple to brown, to 2 cm tall; erect featherlike portion appearing bilaterally branched; main axes with 6-8 transversely divided pericentral cells; branchlets incurved with the terminal segments monosiphonous; extensive creeping filaments below attached by haptera; tetrasporangia borne in stichidia.

This species grows in association with *Caloglossa leprieurii* in brackish water and is particularly abundant during late winter in the Cape Fear estuary near Kure Beach.

Chondria dasyphylla (Woodw.) C. Ag. Figs. 156-162

Plants purple to red, bushy and broadly pyramidal, to 20 cm tall; one-to-few distinct main axes beset with numerous alternate branches; branchlets clustered, club-shaped, basally constricted and flattened at the apex; apical cell sunken and hidden by a dense tuft of trichoblasts; tetrasporangia formed toward the tips of swollen branchlets; pericarps pedicellate, borne in groups of 2-3 on the ultimate branchlets; spermatangial platelets congested near the branchlet tips.

This species is common during late spring at Radio Island and Masonboro Inlet jetties in the upper sublittoral. In the southern part of the state it perennates as decumbent branches, achieving optimal growth and reproductive development in late spring and fall.

Chondria littoralis Harv. Figs. 164-170

Plants straw colored, bushy, to 15 cm tall; one-to-few distinct main axes; denuded below, irregularly and openly branched above with elongate main branches resembling main axes; branchlets elongate, slender, basally constricted; distal tips usually pointed or occasionally flattened, beset with conspicuous persistent trichoblasts; tetrasporangia formed throughout the branchlets; pericarps pedicellate, solitary; spermatangial platelets persisting along the middle of the branchlets.

This species is darker and smaller than *C. dasyphylla*, with markedly more slender branches. Both species occur in the same habitats, perenniate as decumbent branches, and achieve maximum development in spring and fall.

Although *C. baileyana* and *C. sedifolia* have been reported for coastal North Carolina as well (Hoyt 1920; Williams 1948a, 1951; Taylor 1957, 1960), it is probable that most of these records refer to growth forms of *C. dasyphylla* and *C. littoralis* as delimited by Børgesen (1913-20). Taylor (1957, 1960) considered *C. baileyana* and *C. sedifolia* to be primarily northern species and regarded most southern records with doubt. In support of this view, it is unlikely that these species, if present in our flora, would achieve maximum development during the warmer months as reported (Hoyt 1920; Williams 1948a). Furthermore, Hoyt (1920) described the extreme morphological variation among specimens he variously attributed to these taxa and expressed concern that his determinations were not correct.

Chondria polyrhiza Coll. et Herv. Figs. 171-173

Plants minute, creeping, to 5 mm tall, attached to the substratum by bundles of multicellular rhizoids; branchlets with contracted bases and a conspicuous tuft of trichoblasts which do not obscure the apical cell at the pointed tip.

This species, epiphytic on pelagic *Sargassum*, is infrequently collected and was found once on the Masonboro Inlet breakwater. It is previously unreported for North Carolina, but is found in Bermuda and Florida (Taylor 1960). *Chondria curvilineata*, which is similar to *C. polyrhiza*, is reported from offshore (Schneider 1975a).

Chondria tenuissima (Good. et Woodw.) C. Ag. Figs. 174-176

Plants bushy, 10-15 cm tall, light red, soft; main axis

with several similar branches beset with clusters of branchlets; branchlets elongate, to 5 mm, tapering at both ends, the tips pointed, with an exposed apical cell and rather sparse tufts of trichoblasts.

This species was collected once during this study at Oregon Inlet, but is previously reported for Cape Lookout (Williams 1949) and Onslow Bay (Schneider 1976).

Herposiphonia tenella (C. Ag.) Ambr. (= *Herposiphonia secunda* [C. Ag.] Ambr.) Figs. 177-179

Plants minute, dark red epiphytes consisting of extensive prostrate filaments forming entangled mats; apices of main axes upcurved; erect branches arising on the dorsal side, often with a branchlet or branch rudiment from each node; one-to-few branchlets alternating with indeterminate branches; erect branches to 3 mm tall; plants with 8-10 pericentral cells throughout; tetrasporangia swollen, in long series in the middle of the erect branches; antheridial branches arise from the entire trichoblast primordium in series in the branch tips, pericarps nearly spherical to 400 µm diameter, with an elongate neck and contracted ostiole.

This species is a common epiphyte on coarser algae, especially on *Padina* stipes, during the warmer months. Plants with reproductive structures are typically collected in August and September.

Hollenberg (1968a) synonymized *H. secunda* as a varietal form of *H. tenella*, thus *H. tenella* v. *secunda* (C. Ag.) Hollenberg. This taxonomic revision, however, is not universally accepted (Taylor 1960). Morrill (1977), comparing branch sequence patterns and placement of reproductive structures in erect branches, recently demonstrated a basis for distinguishing between these two entities sufficient to warrant retention of specific status for each.

Laurencia corallopsis (Mont.) Howe Figs. 181, 186

Plants rose red, to 15 cm tall, fleshy; one-to-few main axes arising near the holdfast; sparingly, alternately branched above the holdfast; branches blunt-tipped and resembling main axes; peglike branchlets few and inconspicuous; in transection cortical cells square or radially elongate.

This species was previously reported from the continental shelf (Schneider 1976) and in this study was found on near-shore ledges at moderate depth and occasionally in the drift.

Laurencia poitei (Lamour.) Howe Figs. 180, 182-185

Plants yellow brown, cartilaginous, to 20 cm tall; bushy and intricately branched; main axes alternately branched and beset with numerous, pinnate peglike branchlets, mostly about 2 mm long; in transection cortical cells rounded.

This species was occasionally found adrift or dredged in bays and tidal creeks, especially during the warmer months. Like *Gracilaria verrucosa* and *Gigartina acicularis* with which it grows, this alga is probably more common than generally assumed, but its preference for turbid waters and its dark color make it easily overlooked.

Five species of *Laurencia* are reported from North Carolina (Hoyt 1920; Williams 1951; Taylor 1960; Schneider 1976), with most of these records referring to offshore collections.

Systematic Account : 75

According to Woelkerling (1975), American *Laurencia* species from the tropical Atlantic are in need of critical monographic investigation in that specific identifications have apparently been made using characteristics of questionable taxonomic significance. Elsewhere, *Laurencia* species have been critically studied by Saito (1967, 1969) and Jaasund (1970, 1976).

Micropeuce mucronata (Schm.) Kylin (= *Brongniartella mucronata* [Harv.] Schm.) Figs. 139-140, 163

Plants light red, bushy, to 15 cm tall; main axes polysiphonous with 5 pericentral cells and superficially corticated with rhizoidal cells; older axes denuded, branchlets with subdichotomously branched monosiphonous ramelli.

This species is occasionally found in the drift or on near-shore ledges. Hoyt (1920) reported specimens from offshore dredges; it was also found in abundance in an extensive survey of the continental shelf (Schneider 1976).

Polysiphonia

The genus *Polysiphonia* is one of the most widely distributed and diverse taxa in the world. In local waters it is represented by thirteen species, some of which seasonally comprise a large proportion of our coastal eulittoral flora. This genus is the subject of a recent monograph (Kapraun 1977a) that includes detailed descriptions and illustrations for nine species in North Carolina. Four additional species, *P. howei*, *P. subtilissima*, *P. pseudovillum*, and *P. flaccidissima* are included in the present account.

Although specific determinations in this large genus

are often reported to be difficult (Taylor 1960; Edwards 1970), critical attention to morphological details and an awareness of inherent variation in local populations will allow for ready identification of freshly collected or liquid-preserved material. As the key indicates, characteristics assumed to have taxonomic significance include number of pericentral cells, origin of branches (replacing trichoblasts or arising in connection with them), origin of rhizoids (cut off from or in open connection with pericentral cells), and nature of antheridial branches and cystocarps. Information concerning the importance of these characteristics is available elsewhere (Børgesen 1913-20; Hollenberg 1942, 1944, 1958, 1961, 1968b, 1968c; Segi 1951; Kapraun 1977a) and will not be repeated here.

Polysiphonia denudata (Dillw.) Kuetz. Figs. 193-195

Plants red to purple, to 15 cm tall, with a discoidal base, but occasionally with erect axes from decumbent branches; branches arising in axils of trichoblasts; rhizoids cut off from the proximal end of pericentral cells; 5-6 pericentral cells; spermatangial branches long, conical, often with 1-3 sterile tip cells, and subtended by a branched trichoblast; pericarps spherical with a narrow ostiole; tetrasporangia in long series in the upper branches.

This is a species found in bays and sounds, where it achieves maximum development during the winter and spring, but persists in reduced form throughout the year.

Polysiphonia ferulacea Suhr Figs. 219-221

Plants light reddish brown, to 6 cm tall, with a discoidal

base and distinct main axis; branches replace trichoblasts in development; 4 pericentral cells; rhizoids cut off from the end of pericentral cells; spermatangial branches cylindrical with 1-2 sterile tip cells, subtended by a branched trichoblast; cystocarps spherical, opening by a flaring ostiole; tetrasporangia in short series in the branch tips.

This species is locally abundant in the Wrightsville Beach area, which is apparently its northernmost point of distribution on our coast. Local plants are more delicate than the coarse, dark specimens encountered in the tropics, although the reproductive structures are remarkably similar so that identifications can be made with confidence.

A recent cytological investigation of this species demonstrated the existence of genetically isolated populations ($N = 27$, $N = 30$), with the latter utilizing specialized asexual propagules as an accessory mode of reproduction (Kapraun 1977c, 1978b).

Polysiphonia flaccidissima Hollenb. Figs. 205-206

Plants minute, to 1 cm tall, arising from prostrate filaments; branches arising in the axils of trichoblasts; rhizoids cut off from the proximal end of pericentral cells; 4 pericentral cells; tetrasporangia swollen in the upper branches.

Although this species was not encountered during a previous intensive study of *Polysiphonia* in North Carolina (Kapraun 1977a), it has been reported as an epiphyte of *Zostera* in the Beaufort area (Brauner 1975). More recently, vegetative and tetrasporophytic plants were collected as epiphytes on *Sargassum* at Wrightsville Beach. As observed by Brauner (1975), local specimens do not agree completely with the descriptions provided by Hollenberg (1968a) for specimens of the central Pacific, but are in close agreement with his original diagnosis for Pacific coast plants (Hollenberg 1942).

Polysiphonia harveyi Bailey   Figs. 187-189

Plants brown to red, to 10 cm tall, coarse, broadly pyramidal with one-to-few distinct main axes from a discoidal base; branches arising in the axils of branches, but obscurely so; older axes with rhizoidal cortication and adventitious branching from trichoblast scar cells; 4 pericentral cells; spermatangial branches cylindrical to lanceolate, with or without 1-2 sterile tip cells, but subtended by a branched trichoblast; cystocarps variable, spherical to urceolate, opening by a wide ostiole; tetrasporangia swollen in branchlets.

Polysiphonia harveyi is one of the dominant species in our area during winter and early spring when it forms dense mats in the littoral on jetty rocks. A recent cytological investigation of the morphological forms of this species (var. *arietina sensu* Taylor and *v. olneyi*) revealed that these varieties are genetically distinct sibling species (Kapraun 1978b).

Polysiphonia havanensis Mont. *sensu* Børg.   Figs. 199-201

Plants reddish brown, soft, 2-4 (-8) cm tall with an extensive creeping system; branches arising in axils of trichoblasts; 4 pericentral cells; rhizoids developing from the middle of pericentral cells and remaining in open connection

with them; spermatangial branches long, cylindrical, lacking
sterile tip cells, subtended by branching trichoblasts; cystocarps broadly urn-shaped; tetrasporangia in long series
throughout the main branches.

As previously discussed (Kapraun 1977a), this entity is
probably synonymous with *P. sertularioides* from the Mediterranean. Lauret* examined American material of *P. havanensis
sensu* Børgesen and found no substantial difference between
it and *P. sertularioides*, which he has studied critically
(Lauret 1967).

Polysiphonia howei Hollenb. Figs. 202-204

Plants minute, to 1 cm tall from prostrate filaments
with upward bent tips; branches in erect filaments replacing
trichoblasts in development; 8-10 pericentral cells; trichoblasts highly branched, with a characteristically short basal
cell below a comparatively large, rounded branch cell, rhizoids cut off from the distal end of pericentral cells; erect
branches arise at distant intervals exogenously at the tips
of prostrate branches, or adventitiously from scar cells in
prostrate branches, pericentral cells shifting to offset
positions and forming secondary pit connections with the 2
pericentral cells of adjacent segments.

Although this species was not encountered during a previous intensive study of *Polysiphonia* in North Carolina, it
has been reported from offshore (Williams 1951; Hollenberg
1958). Recently, vegetative plants were collected as epiphytes on *Sargassum* at Wrightsville Beach. These specimens

---

*M. Lauret 1979: personal communication.

closely resembled the descriptions of this species in the Pacific (Hollenberg 1968b).

Polysiphonia macrocarpa Harv. Figs. 196-198

Plants minute, reddish black, forming dense tufts 1-2 cm tall; branches replacing trichoblasts in development; 4 pericentral cells; rhizoids developing from the middle of pericentral cells and remaining in open connection with them; prostrate filaments with upward bent tips devoid of trichoblasts, giving rise to adventitious branches at intervals of 4-6 segments; spermatangial branches long, cylindrical, arising from the entire trichoblast primordium; cystocarps urceolate and large (to 250 μm diameter) relative to the minute size of the plants; tetrasporangia swollen in long series in the tips of erect filaments.

The habit and seasonality of this species and *P. howei* are similar, but the latter, with 8-10 pericentral cells and trichoblasts on prostrate branches, is easily distinguished.

Polysiphonia nigrescens (Huds.) Grev. Figs. 215, 218

Plants dark, coarse, erect, to 25 cm from a discoid base; alternately branched from one-to-few main axes to 600 μm diameter and ultimately corticated below; adventitious branches abundant; branches replacing trichoblasts in development; 10-14 pericentral cells; cystocarps broadly oval with a narrow ostiole; tetrasporangia in long series in forked branched tips.

This is the largest and coarsest *Polysiphonia* species in our area. It is uncommon, with individual plants being collected at Beaufort and Wrightsville Beach during winter and early spring.

Polysiphonia pseudovillum Hollenb.   Figs. 207-210

Plants minute, creeping, erect filaments to 2 mm tall; 4 pericentral cells; rhizoids cut off from the proximal end of pericentral cells; branches arising cicatrigenously at intervals of 4-8 segments from prostrate axes; erect filaments with persistent, slender trichoblasts; tetrasporangia little swollen in spiral arrangement in branch tips.

This species was found epiphytic on pelagic *Sargassum* at Wrightsville Beach and is previously unreported for our coast.

Along with *Polysiphonia macrocarpa* and *P. flaccidissima*, *P. pseudovillum* is a minute epiphyte with an extensive prostrate system. It differs from the former, which has rhizoids remaining in open connection with pericentral cells, and from the latter, which has branches arising in the axils of trichoblasts.

Polysiphonia sphaerocarpa Børg.   Figs. 190-192

Plants small epiphytes, less than 2 cm tall from a discoidal base, branching dichotomous; branches becoming decumbent and attaching to the substratum by rhizoids cut off from pericentral cells; branches replacing trichoblasts in development; 4 pericentral cells; spermatangial branches cylindrical, with or without sterile tip cells and subtended by a branched trichoblast; cystocarps spherical and remaining so even after the large-celled ostiolar rim opens; tetrasporangia in short series in forked branch tips.

This minute plant is a common epiphyte on coarser algae during summer, but its small size makes it easily overlooked. Contrary to previous descriptions (Taylor 1960), this species does not have a base of creeping filaments as does *P. macrocarpa* and *P. howei*. Bearing this fact in mind and using the description above, *P. sphaerocarpa* in our area can be easily identified.

Polysiphonia subtilissima Mont. Figs. 222-225

Plants olive to purple, soft, forming dense epiphytic tufts 2-4 cm tall; branches replacing trichoblasts in development; 4 pericentral cells; erect and prostrate apices with similar radial development of branch initials; apical cells conspicuous, large, 10 x 15 µm, trichoblasts scarce, little branched; scar cells giving rise to adventitious branches; erect filaments repeatedly dichotomously branched, 50-60 µm diameter; prostrate filaments 80-120 µm; rhizoids remaining in open connection with pericentral cells; reproductive structures not seen.

This species was not encountered in a previous study of *Polysiphonia* in North Carolina (Kapraun 1977a), but was reported for the state by Taylor (1960). Material included herein was collected as epiphytes on *Spartina* in brackish water during November, along with *Caloglossa leprieurii* and *Bostrychia radicans*. As suggested by Taylor (1960), this species is probably more common than published accounts indicate, but its small size, dark color, and restriction to muddy, brackish habitats make it easily overlooked.

Along with *P. macrocarpa* and *P. urceolata*, *P. subtilissima* is a species with 4 pericentral cells, rhizoids remaining in open connection with pericentral cells, and an extensive prostrate system which gives rise to erect filaments. Despite these similarities, these taxa are readily distinguished.

Both *P. subtilissima* and *P. urceolata* have radial development of branches in both erect and upturned prostrate apices. Although erect filaments of *P. macrocarpa* have radial organization, in contrast it possesses prostrate axes which give rise to unilateral filaments, giving plants a dorsiventral habit. Hollenberg (1968b) discussed how dorsiventrality in such cases is secondarily derived from erect branches arising cicatrigenously from primordia resembling scar cells which are in fact spirally arranged. Critical attention to this character easily distinguishes *P. macrocarpa* and *P. subtilissima*, which are small, dark plants. *P. urceolata*, a much larger, lighter colored plant, would scarcely be confused with either of the above. In addition, *P. subtilissima* is restricted to brackish habitats and has no known sexual reproductive structures, but *P. urceolata* and *P. macrocarpa* are found only on the open coast and have distinctive antheridial branches and pericarps.

Polysiphonia tepida Hollenb. Figs. 216-217

Plants dark red, forming mats 1-2 cm (to 10 cm) tall; extensive prostrate filaments attached to the substratum by rhizoids cut off from pericentral cells; branches arising in axils of trichoblasts; 7-8 pericentral cells; spermatangial branches elongate, cylindrical, often bifurcate, and subtended by a branched trichoblast; cystocarps urceolate; tetrasporangia swollen, in short series in the middle of erect branches.

This species is locally abundant in sand at the Radio Island and Masonboro Inlet jetties during April and May, when plants become 10 cm tall before dying back to shorter entangled mats which persist through summer.

Polysiphonia urceolata (Lightf.) Grev. Figs. 211-214

Plants yellow brown to light red, to 10 cm tall from an extensive creeping base; prostrate filaments attached to the substratum by rhizoids remaining in open connection with pericentral cells; branches replacing trichoblasts in development; 4 pericentral cells; spermatangial branches long, cylindrical, usually without sterile tip cells, and subtended by a branched trichoblast; cystocarps broadly urceolate, but lacking an elongate neck and constricted ostiole; tetrasporangia swollen in long series in lateral branches.

Comparative morphological and cultural studies of North American and Norwegian populations of this species have demonstrated considerable variation in reproductive structures and growth responses, suggesting genetically based ecotypic variation in this widely distributed taxon (Kapraun 1979).

Pterosiphonia pennata (Roth) Falkenb. Fig. 227

Plants solitary, dark reddish purple, to 2 cm tall; irregularly pinnately branched; main axes with 8-9 pericentral cells, becoming lightly corticated below; branchlets initially incurved, simple or sparingly rebranched.

This minute plant is occasionally encountered at low water mark on the Radio Island jetty during summer and has been reported from Cape Lookout (Williams 1948a) as well as from coastal South Carolina (Wiseman and Schneider 1976).

Pterosiphonia pennata and *Bostrychia radicans* are superficially similar in that they have about eight pericentral cells and lack trichoblasts. Consequently, they have been confused on our coast (Blomquist and Humm 1946). However, *Pterosiphonia* plants have a regular system of equal length pinnate branches, giving a featherlike appearance, while *Bostrychia* has alternate branches of increasing length toward the base, giving a pyramidal habit. In addition, *Bostrychia* branchlets are monosiphonous, incurved toward the tips, and tend to rebranch in one-to-few orders, while *Pterosiphonia* branchlets are completely polysiphonous, straight, and simple.

Literature Cited

Abbott, I. A.; Hollenberg, G. J. 1976. *Marine Algae of California*. Stanford: Stanford University Press. 827 pp.

Ardré, F. 1970a. Contribution à l'étude des algues marines du Portugal: I, La Flore. *Portugaliae Acta Biologica* (B), 10, 423 pp.

―――――. 1970b. Contribution à l'étude des algues marines du Portugal: II, Ecologie et Chorologie. *Bull. Cent. Etud. Rech. Scient. Biarritz.* 8:359-574.

Aregood, C. 1975. A study of the red alga *Calonitophyllum medium* (Hoyt) comb. nov. (= *Hymenena media* [Hoyt] Taylor). *Br. Phyc. J.* 10:347-62.

Aziz, K. M. S.; Humm, H. J. 1962. Additions to the algal flora of Beaufort, North Carolina, and vicinity. *J. Elisha Mitchell Sci. Soc.* 78:55-63.

Baldock, R. N. 1976. The Griffithsiae group of the Ceramiaceae (Rhodophyta) and its southern Australian representatives. *Aust. J. Bot.* 24:509-93.

Blomquist, H. L.; Humm, H. J. 1946. Some marine algae new to Beaufort, North Carolina. *J. Elisha Mitchell Sci. Soc.* 68:1-8.

Børgesen, F. 1913-20. The marine algae of the Danish West Indies: I, Chlorophyceae. *Dansk. Bot Arkiv.* 1:1-158, 1913; II, Phaeophyceae, 2:1-66, 1914; III, Rhodophyceae, 3:1-504, 1920.

Brauner, J. F. 1975. Seasonality of epiphytic algae on
Zostera marina at Beaufort, North Carolina. *Nova Hedwigia* 26:125-33.

Chapman, A. R. O.; Edelstein, T.; Power, P. J. 1977. Studies on *Gracilaria*: I, Morphological and anatomical variation in samples from the lower Gulf of St. Lawrence and New England. *Bot. Mar.* 20:149-53.

Coll, J.; Cox, J. 1977. The genus *Porphyra* C. Ag. (Rhodophyta, Bangiales) in the American North Atlantic: I, New species from North Carolina. *Bot. Mar.* 20:155-60.

Collins, F. S. 1903. The Ulvaceae of North America. *Rhodora* 5:1-31.

_____. 1909. The Green Algae of North America. *Tufts Coll. Stud. Sci. Ser.* 2:79-480.

_____; Hervey, A. B. 1917. The Algae of Bermuda. *Proc. Am. Acad. Arts and Sci.* 53:1-195.

Conover, J. T. 1958. Seasonal growth of benthic marine plants as related to environmental factors in an estuary. *Publ. Inst. Mar. Sci. Univ. Tex.* 5:97-147.

Conway, E.; Mumford, T.; Scagel, R. 1975. The genus *Porphyra* in British Columbia and Washington. *Syesis* 8:185-244.

Dawes, C. J. 1974. *Marine algae of the west coast of Florida*. Coral Gables: University of Miami Press. 201 pp.

Dawson, E. Y. 1941. Review of the genus *Rhodymenia* with descriptions of new species. *Allan Hancock Pacific Expeditions*, vol. 3, nos. 7-8, pp. 115-56. Los Angeles: University of Southern California Press.

_____. 1966a. *Marine Botany: An Introduction*. New York: Holt, Rinehart, & Winston. 371 pp.

_____. 1966b. *Seashore Plants of Southern California*. Berkeley: University of California Press. 101 pp.

Dixon, P. 1958. The structure and development of the thallus in the British species of *Gelidium* and *Pterocladia*. *Ann. Bot.* 22:353-68.

_____. 1961. On the classification of the Florideae with particular reference to the position of the Gelidiaceae. *Bot. Mar.* 3:1-16.

_____. 1963. Variation and speciation in marine Rhodophyta. *Oceanogr. Mar. Biol. Ann. Rev.* 1:177-96.

_____. 1966. On the form of the thallus in the Florideophyceae. In: E. Cutter (ed.), *Trends in plant morphogenesis*, pp. 45-63. London: Longmans, Green.

_____. 1970. The Rhodophyta: some aspects of their biology, II. *Oceanogr. Mar. Biol. Ann. Rev.* 8:307-52.

_____. 1973. *Biology of the Rhodophyta*. New York: Hafner Press. 285 pp.

_____; Irvine, L. 1977. Seaweeds of the British Isles: I, Rhodophyta. Part I. Introduction, Nemaliales, Gigartinales. London: British Museum (N.H.). 252 pp.

D'Lacoste, L. E.; Ganesan, E. K. 1972. A new freshwater species of *Rhodochorton* (Rhodophyta, Nemaliales) from Venezuela. *Phycologia* 11:233-38.

Earle, L. E.; Humm, H. J. 1964. Intertidal zonation of algae in Beaufort Harbor. *J. Elisha Mitchell Sci. Soc.* 80:78-82.

Edelstein, T. 1977. Studies on *Gracilaria* sp.: Experiments on inocula incubated under greenhouse conditions. *J. Exp. Mar. Biol. Ecol.* 30:249-59.

Literature Cited : 90

_____; Chen, L. C.-M.; McLachlan, J. 1978. Studies on *Gracilaria* (Gigartinales, Rhodophyta): reproductive structures. *J. Phycol.* 14:92-100.

Edwards, P. 1969. Field and cultural studies on the seasonal periodicity of growth and reproduction of selected Texas benthic marine algae. *Contr. Mar. Sci. Univ. Tex.* 14:59-114.

_____. 1970. Illustrated guide to the seaweeds and sea grasses in the vicinity of Port Aransas, Texas. *Contr. Mar. Sci. Univ. Tex. (suppl.).* 128 pp.

_____. 1971. The effects of light intensity, daylength and temperature on the growth and reproduction of *Callithamnion byssoides*. In: B. Parker and M. Brown (eds.), *Selected papers in phycology*, pp. 163-73. Lawrence, Kans.: Allen Press.

_____; Kapraun, D. 1973. Benthic marine algal ecology in the Port Aransas, Texas area. *Contr. Mar. Sci. Univ. Tex.* 17:15-52.

Farlow, W. G. 1881. *The Marine Algae of New England.* Rep. U.S. Fish and Fisheries for 1879, Appendix A-1:1-210.

Feldmann-Mazoyer, G. 1941. *Recherches sur les Ceramiacées de la Méditerranée Occidentale.* Alger. 510 pp.

Fiore, J. 1969. Life-history studies of Phaeophyta from the Atlantic coast of the United States. Ph.D. thesis, Duke University. 242 pp.

_____. 1977. Life history and taxonomy of *Stictyosiphon subsimplex* Holden (Phaeophyta, Dictyosiphonales) and *Farlowiella onusta* (Kützing) Kornmann in Kuckuck (Phaeophyta, Ectocarpales). *Phycologia* 16:301-11.

Literature Cited : 91

Garbary, D. J.; Grund, D.; McLachlan, J. 1978. The taxonomic status of *Ceramium rubrum* (Huds.) C. Ag. (Ceramiales, Rhodophyceae) based on culture experiments. *Phycologia* 17:85-94.

Gayral, P. 1958. *Algues de la côte Atlantique morocaine.* La nature au Maroc, II. Rabat. 523 pp.

_____. 1966. *Les Algues de côtes françaises.* Paris: Editions Doin. 632 pp.

Geesink, R. 1973. Experimental investigations on marine and freshwater *Bangia* (Rhodophyta) from the Netherlands. *J. Exp. Mar. Biol. Ecol.* 11:239-47.

Guiry, M. D. 1974a. A reappraisal of the genus *Palmaria* Stackhouse. *Br. Phycol. J.* 9:219 (abstract).

_____. 1974b. A preliminary consideration of the taxonomic position of *Palmaria palmata* (L.) Stackh. (= *Rhodymenia palmata* [L.] Grev.) *J. Mar. Biol. Assoc. U. K.* 54:509-28.

Harvey, W. H. 1852-58. Nereis Boreali-Americana. I, Melanospermae. *Smithsonian Contrib. to Knowledge,* 3(4):1-150, 1852; II, Rhodospermae, 5(5):1-258, 1853; III, Chlorospermae, 10:1-140, 1858.

Hine, A. E. 1976. *RSMAS Technical Report: A glossary of phycological terms for students of marine macroalgae.* Coral Gables: University of Miami Press. 80 pp.

Hoek, C. van den. 1975. Phytogeographic provinces along the coasts of the Northern Atlantic Ocean. *Phycologia* 14:317-30.

_____; Cortel-Breeman, A. M. 1970. Life-history studies on Rhodophyceae: III, *Scinaia complanata* (Collins) Cotton. *Acta Bot. Neerl.* 19:457-67.

Hollenberg, G. 1942. An account of the species of *Polysiphonia* on the Pacific coast of North America: I, Oligosiphonia. *Am. J. Bot.* 29:772-85.

_____. 1944. An account of the species of *Polysiphonia* on the Pacific coast of North America: II, Polysiphonia. *Am. J. Bot.* 31:474-83.

_____. 1958. Phycological Notes, II. *Bull. Torrey Bot. Club* 85:63-69.

_____. 1961. Marine red algae of Pacific Mexico: V, The genus *Polysiphonia*. *Pac. Nat.* 2:345-75.

_____. 1968a. An account of the species of the red alga *Herposiphonia* occurring in the central and western tropical Pacific Ocean. *Pac. Sci.* 22:536-59.

_____. 1968b. An account of the species of *Polysiphonia* of the central and western tropical Pacific Ocean: I, Oligosiphonia. *Pac. Sci.* 22:56-98.

_____. 1968c. An account of the species of *Polysiphonia* of the central and western tropical Pacific Ocean: II, Polysiphonia. *Pac. Sci.* 22:198-207.

Howe, M. A. 1918. Algae. In: N. L. Britton, *Flora of Bermuda*, pp. 489-540. New York: C. Scribner's Sons.

_____. 1920. Algae. In: N. L. Britton and C. F. Millspaugh, *The Bahama Flora*, pp. 553-618. New York: G. K. Ackerman.

Hoyt, W. D. 1920. Marine algae of Beaufort, North Carolina and adjacent regions. *Bull. Bur. Fish. (U.S.)* 36:372-550.

Humm, H. 1969. Distribution of marine algae along the Atlantic coast of North America. *Phycologia* 7:43-53.

Jaasund, E. 1970. Marine algae in Tanzania, II. *Bot. Mar.* 13:59-64.

_____. 1976. *Intertidal Seaweeds in Tanzania.* Norway: University of Tromsø Press. 160 pp.

Johansen, H. W. 1974. Articulated coralline algae. *Oceanogr. Mar. Biol. Ann. Rev.* 12:77-127.

_____. 1976. Current status of generic concepts in coralline algae (Rhodophyta). *Phycologia* 15:221-44.

Kapraun, D. F. 1974. Seasonal periodicity and spatial distribution of benthic marine algae in Louisiana. *Contr. Mar. Sci. Univ. Tex.* 18:139-67.

_____. 1977a. The genus *Polysiphonia* in North Carolina, USA. *Bot. Mar.* 20:313-31.

_____. 1977b. Studies on growth and reproduction of *Antithamnion cruciatum* (Rhodophyta, Ceramiales) in North Carolina. *Norw. J. Bot.* 24:269-74.

_____. 1977c. Asexual propagules in the life history of *Polysiphonia ferulacea* (Rhodophyta, Ceramiales) from North Carolina. *Phycologia* 16:417-26.

_____. 1978a. Field and culture studies on growth and reproduction of *Callithamnion byssoides* (Rhodophyta, Ceramiales) in North Carolina. *J. Phycol.* 14:21-24.

_____. 1978b. A cytological study of varietal forms in two species of *Polysiphonia* (Rhodophyta, Ceramiales) in North Carolina. *Phycologia* 17:152-56.

_____. 1978c. Field and culture studies on selected North Carolina *Polysiphonia* species. *Bot. Mar.* 21:143-53.

_____. 1979. Comparative studies of *Polysiphonia urceolata* (Lightfoot) Greville (Ceramiales, Rhodophyta) from

Literature Cited : 94

the North Atlantic. *Norw. J. Bot.* 26:269-76.

Kormann, P.; Sahling, P. 1977. Meeresalgen von Helgoland: Bentische Grün-, Braun-, und Rotalgen. *Helgolander wiss. Meeresunters* 19. 165 pp.

Kurogi, M. 1972. Systematics of *Porphyra* in Japan. In: I. A. Abbott and M. Kurogi (eds.), *Contributions to the Systematics of Benthic Marine Algae of the North Pacific*, pp. 167-92. Kobe, Japan: Japanese Society of Phycology.

Lauret, M. 1967. Morphologie, phénologie, répartition des *Polysiphonia* marins du littoral languedocien: I, section Oligosiphonia. *Naturalia monspeliensia, série Bot.* 18:347-73.

Littler, M. M. 1972. The crustose Corallinaceae. *Oceanogr. Mar. Biol. Ann. Rev.* 10:311-47.

MaLachlan, J. 1979. *Gracilaria tikvahiae* sp. nov. (Rhodophyta, Gigartinales, Gracilariaceae) from the northwestern Atlantic. *Phycologia* 18:19-23.

Morrill, J. F. 1977. Graphic representation of morphological variation among species of certain dorsiventral Rhodomelaceae. *J. Phycol.* 13(suppl.):46.

Nichols, H.; Lissant, E. 1967. Developmental studies of *Erythrocladia* Rosenvinge in culture. *J. Phycol.* 3:6-18.

Papenfuss, G. F. 1961. The structure and reproduction of *Caloglossa leprieurii*. *Phycologia* 1:8-31.

_____. 1968. Notes on South African marine algae. V. *J. S. Afr. Bot.* 24:267-87.

Phillips, R. C. 1961. Seasonal aspect of the marine flora of the St. Lucie inlet and adjacent Indian River, Florida. *Q. J. Fla. Acad. Sci.* 24:135-47.

Literature Cited : 95

Ramus, J. 1969. The developmental sequence of the marine red alga *Pseudogloiophloea* in culture. *Univ. Calif. Publ. Bot.* 52:1-42.

Rhyne, C. 1973. Field and experimental studies on the systematics and ecology of Ulva curvata and Ulva rotundata. UNC-Sea Grant Publ. 73-09. 123 pp.

Richardson, J. 1979. Overwintering of *Dictyota dichotoma* (Phaeophyceae) near its northern distribution limit on the east coast of North America. *J. Phycol.* 15:17-21.

Richardson, N.; Dixon, P. 1968. Life-history of *Bangia fuscopurpurea* (Dillw.) Lyngb. in culture. *Nature, Lond.* 218:496-97.

Rueness, J. 1973. Culture and field observations on growth and reproduction of *Ceramium strictum* Harv. from the Oslo fjord, Norway. *Norw. J. Bot.* 20:61-65.

_____. 1977. *Norsk Algenflora*. Oslo: University of Oslo Press. 266 pp.

_____. 1978. Hybridization in red algae. In: D. E. G. Irvine and J. H. Price (eds.), *Modern Approaches to the Taxonomy of Red and Brown Algae*, pp. 247-62, Systematics Association Special Volume No. 10. London and New York: Academic Press.

Saito, Y. 1967. Studies on Japanese species of *Laurencia*, with special reference to their comparative morphology. *Mem. Fac. Fish. Hokkaido Univ.* 15:1-81.

_____. 1969. The algal genus *Laurencia* from the Hawaiian Islands, the Phillipines and adjacent areas. *Pac. Sci.* 23:148-60.

Santilices, B. 1976. Taxonomic and nomenclatural notes on some Gelidiales (Rhodophyta). *Phycologia* 15:165-73.

Scagel, R. F. 1967. *Guide to Common Seaweeds of British Columbia*. Victoria, B. C.: A. Sutton. 330 pp.

Schneider, C. W. 1974. North Carolina marine algae: III, A community of Ceramiales (Rhodophyta) on a glass sponge from 60 meters. *Bull. Mar. Sci.* 24:1093-1101.

_____. 1975a. North Carolina marine algae: V, Additions to the flora of Onslow Bay, including the reassignment of *Fauchea peltata* Taylor to *Weberella* Schmitz. *Br. phycol. J.* 10:129-38.

_____. 1975b. North Carolina marine algae: VI, Some Ceramiales (Rhodophyta), including a new species of *Dipterosiphonia*. *J. Phycol.* 11:391-96.

_____. 1975c. Taxonomic notes on *Gracilaria mammillaris* (Mont.) Howe and *Gracilaria veleroae* Dawson (Rhodophyta, Gigartinales). *Taxon* 24:643-46.

_____. 1976. Spatial and temporal distributions of benthic marine algae on the continental shelf of the Carolinas. *Bull. Mar. Sci.* 26:133-51.

_____; Searles, R. B. 1973. North Carolina marine algae: II, New records and observations of the benthic offshore flora. *Phycologia* 12:201-11.

_____; Searles, R. B. 1975. North Carolina marine algae: IV, Further contributions from the continental shelf, including two new species of Rhodophyta. *Nova Hedwigia* 26:83-103.

_____; Searles, R. B. 1976. North Carolina marine algae: VII, New species of *Hypnea* and *Petroglossum* (Rhodophyta, Gigartinales) and additional records of other Rhodophyta. *Phycologia* 15:51-60.

Literature Cited : 97

Searles, R. B. 1972. North Carolina marine algae: I, Three new species from the continental shelf. *Phycologia* 11:19-24.

_____.; Schneider, C. W. 1978. A checklist and bibliography of North Carolina Seaweeds. *Bot. Mar.* 21:99-108.

Segi, T. 1951. Systematic study of the genus *Polysiphonia* from Japan and its vicinity. *J. Fac. Fish., Mie Prefecture Univ.* 1:169-272.

Stegenga, H.; Borsje, W. J. 1976. The morphology and life history of *Acrochaetium dasyae* Collins (Rhodophyta, Nemaliales). *Acta Bot. Neerl.* 24:15-29.

_____; Vroman, M. 1976. The morphology and life history of *Acrochaetium densum* (Drew) Papenfuss (Rhodophyta, Nemaliales). *Acta Bot. Neerl.* 25:257-80.

Stephenson, T. A.; Stephenson, A. 1952. Life between tide marks in North America: II, Northern Florida and the Carolinas. *J. Ecol.* 40:1-49.

Stewart, J. G. 1968. Morphological variations in *Pterocladia pyramidale. J. Phycol.* 4:76-84.

Stosch, H. A. von. 1965. The sporophyte of *Liagora farinosa* Lamour. *Br. Phycol. Bull.* 2:486-96.

Taylor, W. R. 1928. The marine algae of Florida, with special reference to the Dry Tortugas. Papers from the Tortugas Laboratory, 25:1-219; Carnegie Institute of Washington, Publication 379.

_____. 1957. *Marine algae of the Northeastern Coast of North America.* 2nd ed. Ann Arbor: University of Michigan Press. 509 pp.

Literature Cited : 98

_____. 1960. *Marine Algae of the Eastern Tropical and Subtropical Coasts of the Americas*. Ann Arbor: University of Michigan Press. 870 pp.

Umezaki, I. 1967. The tetrasporophyte of *Nemalion vermiculare* Sur. *Rev. Algol.*, n.s. 9:19-24.

West, J. A. 1968. Morphology and reproduction of the red alga *Acrochaetium pectinatum* in culture. *J. Phycol.* 4:89-99.

Whittick, A.; Hooper, R. G. 1977. The reproduction and phenology of *Antithamnion cruciatum* (Rhodophyta: Ceramiaceae) in insular Newfoundland. *Can. J. Bot.* 55:520-24.

Williams, L. 1948a. Seasonal alternation of marine floras at Cape Lookout, North Carolina. *Am. J. Bot.* 35:682-95.

_____. 1948b. The genus *Codium* in North Carolina. *J. Elisha Mitchell Sci. Soc.* 64:107-16.

_____. 1949. Marine algal ecology at Cape Lookout, North Carolina. *Bull. Furman Univ.* 31:1-21.

_____. 1951. Algae of the Black Rocks. In: A. B. Pearse and L. G. Williams (eds.), *The biota of the reefs off the Carolinas. J. Elisha Mitchell Sci. Soc.* 67:133-61.

Wiseman, D. R. 1966. A preliminary survey of the Rhodophyta of South Carolina. Ph.D. thesis, Duke University. 190 pp.

_____; Schneider, D. W. 1976. Investigations of the marine algae of South Carolina: I. New records of Rhodophyta. *Rhodora* 78:516-24.

Woelkerling, W. J. 1971. Morphology and taxonomy of the
  Audouinella complex (Rhodophyta) in Southern Australia.
  Aust. J. Bot., Suppl. Ser. 1:1-91.

_____. 1972a. Some algal invaders on the northeastern
  fringes of the Sargasso Sea. *Rhodora* 74:295-98.

_____. 1972b. Studies on the *Audouinella microscopica*
  (Naeg.) Woelk. complex (Rhodophyta). *Rhodora* 74:85-96.

_____. 1973a. The morphology and systematics of the
  Audouinella complex (Acrochaetiaceae, Rhodophyta) in
  northeastern United States. *Rhodora* 75:529-621.

_____. 1973b. The *Audouinella* complex (Rhodophyta) in
  the western Sargasso Sea. *Rhodora* 75:78-101.

_____. 1975. On the epibiotic and pelagic Chlorophyceae,
  Phaeophyceae, and Rhodophyceae of the western Sargasso
  Sea. *Rhodora* 77:1-40.

_____. 1976. *South Florida Benthic Marine Algae*. Coral
  Gables: University of Miami Press. 148 pp.

Womersley, H. B. S. 1977. Species concepts in *Ceramium*.
  J. Phycol. 13(suppl.):75.

Wynne, M.; Taylor, W. R. 1973. The status of *Agardhiella
  tenera* and *Agardhiella baileyi* (Rhodophyta, Gigartinales). *Hydrobiologia* 43:93-107.

Illustrations

Figure 1. Map of the study area. Numbers refer to collection sites as follows: 1. Cape Lookout jetty, 2. Fort Macon jetty, 3. Radio Island jetty, Beaufort, 4. Sneads Ferry, New River estuary, 5. Topsail Island Sound, 6. Masonboro Inlet jetty and adjacent sound, Wrightsville Beach, 7. coquina outcroppings at Fort Fisher near Kure Beach, 8. Federal Basin at Fort Fisher, 9. Cape Fear River estuary, 10. Lockwood's Folly Inlet.

Figure 2. *Erythrocladia subintegra.* Scale = 100 μm.

Figure 3. *Goniotrichum alsidii.* Scale = 100 μm.

Figure 4. *Erythrotrichia ciliaris* showing development of pluriseriate filament. Scale = 100 μm.

Figure 5. *Porphyra rosengurtii.* Right: surface view of α, β-spore formation; left: transection of α-spore formation. Scale = 50 μm.

Figures 6-8. *Erythrotrichia carnea.* Fig. 6. Young filament. Fig. 7. Monospore production. Fig. 8. Monospore release. Note highly lobed plastid with a central pyrenoid. Scale = 100 μm.

Figure 9. *Porphyra carolinensis.* Surface view showing dentate margin. Scale = 100 μm.

Figure 10. *Bangia atropurpurea* showing development of pluriseriate filament. Scale = 100 μm.

Illustrations : 103

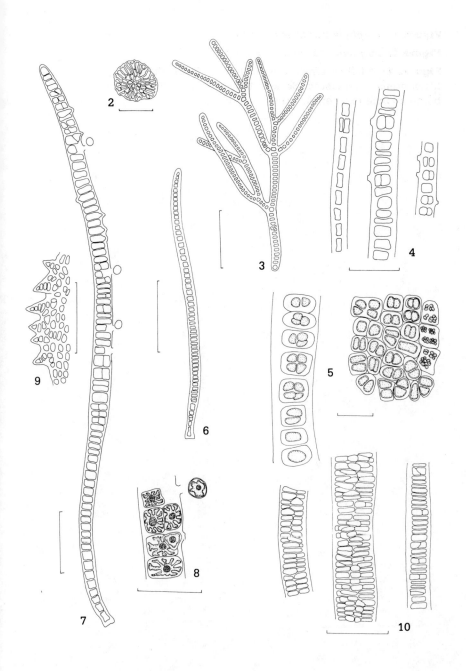

Figure 11. *Porphyra carolinensis.* Scale = 5 cm.
Figure 12. *Porphyra rosengurtii.* Scale = 5 cm.
Figures 13-14. *Audouinella dasyae.* Fig. 13. Habit showing filamentous basal system. Scale = 100 μm. Fig. 14. Monospores borne in unilateral series. Scale = 50 μm.

Illustrations : 105

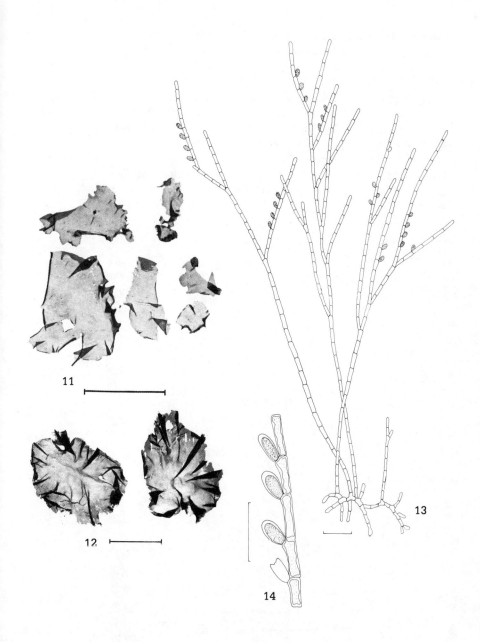

Figures 15-17. *Audouinella alariae.* Fig. 15. Habit showing secund pattern of branching. Scale = 100 μm. Fig. 16. Persistent swollen basal cells. Fig. 17. Monosporangia borne individually or in pairs on stalks. Fig. 16—17. Scale = 50 μm.

Figures 18-20. *Audouinella hallandica.* Fig. 18. Habit showing sparse, pseudodichotomous branching. Scale = 100 μm. Fig. 19. Persistent cylindrical basal cell. Fig. 20. Monosporangia borne in series, individually or in pairs on stalks. Figures 19—20. Scale = 50 μm.

Figures 21-22. *Audouinella microscopica.* Habit showing arcuate branching and monosporangia in series. Scale = 50 μm.

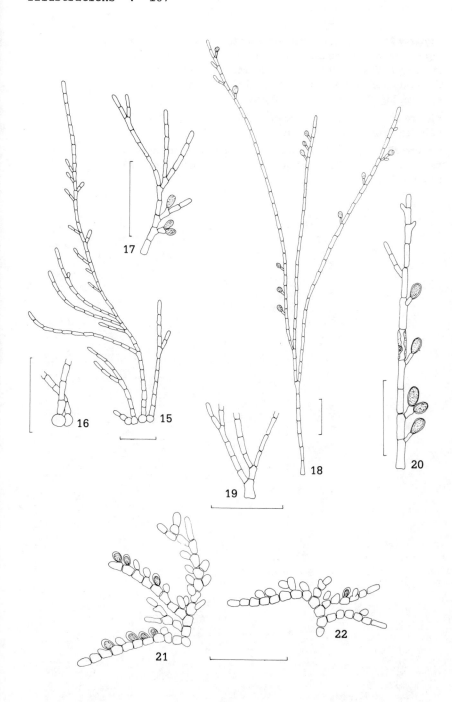

Figures 23-25. *Audouinella secundata.* Fig. 23. Habit showing sparse branching and irregular arrangement of monosporangia. Scale = 100 µm. Fig. 24. Monosporangia borne 2—3 on lateral stalks. Fig. 25. Detail of parenchymatous basal system. Fig. 24—25. Scale = 50 µm.

Figure 26. *Audouinella hallandica.* Monosporangia borne in pairs on stalks. Scale = 100 µm.

Figure 27. *Audouinella thuretii.* Habit showing extensive pseudoparenchymatous basal system and erect filaments. Scale = 50 µm.

Illustrations : 109

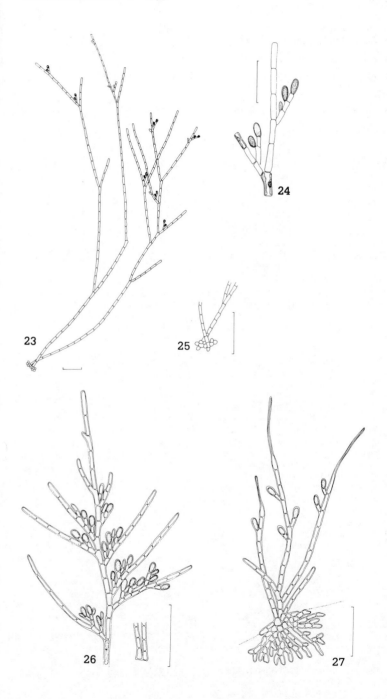

Figure 28. *Scinaia complanata.* Scale = 5 cm.
Figure 29. *Gelidium americanum.* Scale = 2 cm.
Figure 30. *Gelidium crinale.* Plants from sheltered bay. Scale = 2 cm.
Figure 31. *Gelidium crinale.* Plants from surf-exposed rocks. Scale = 5 cm.

Illustrations : 111

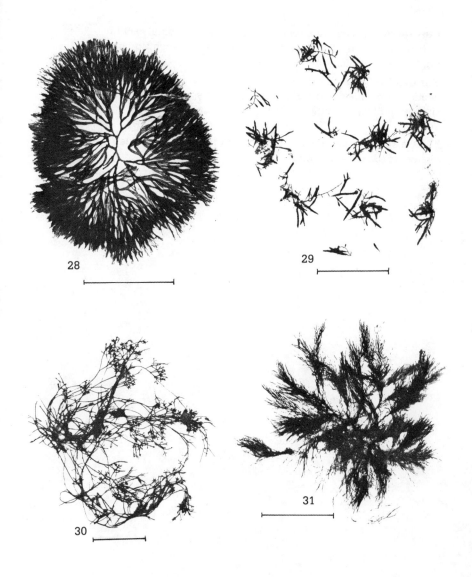

Figures 32-33. *Gelidium crinale.* Fig. 32. Longitudinal section of main axis. Fig. 33. Transection of spatulate branch tip bearing tetrasporangia. Scale = 100 μm.

Figures 34-35. *Gelidium americanum.* Fig. 32. Longitudinal section of main axis. Fig. 33. Transection of spatulate branch tip bearing tetrasporangia. Note cortical rhizines. Scale = 100 μm.

Figure 36. *Gelidium crinale.* Transection through a bilocular cystocarp. Scale = 100 μm.

Illustrations : 113

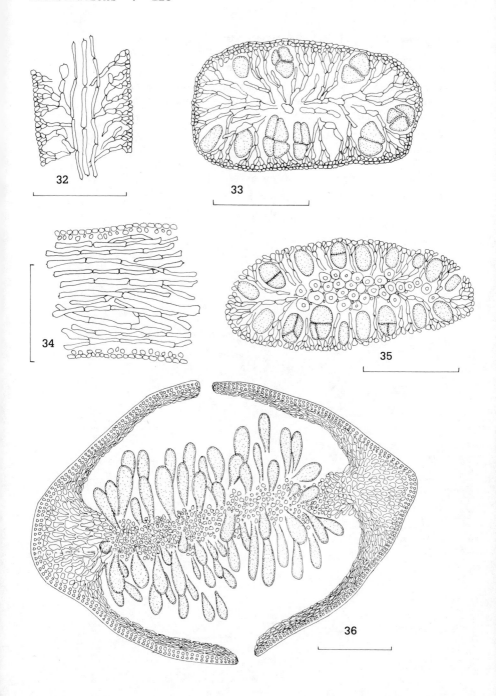

Figures 37-38. *Heteroderma lejolisii.* Fig. 37. Transection of monostromatic vegetative disc. Fig. 38. Transection of cystocarpic conceptacle. Scale = 100 µm.

Figure 39. *Dermatolithon pustulatum.* Transection showing layers of vertically elongate cells and tetrasporangial conceptacles. Scale = 100 µm.

Figures 40-43. *Peyssonellia rubra.* Fig. 40. Transection of scattered cystocarps. Fig. 41. Surface view of vegetative margin. Fig. 42. Transection of tetrasporangial region. Fig. 43. Transection of vegetative margin. Scale = 100 µm.

Figure 44. *Jania adhaerens.* Habit. Scale = 100 µm.

Figure 45. *Corallina cubensis.* Habit. Scale = 100 µm.

Illustrations : 115

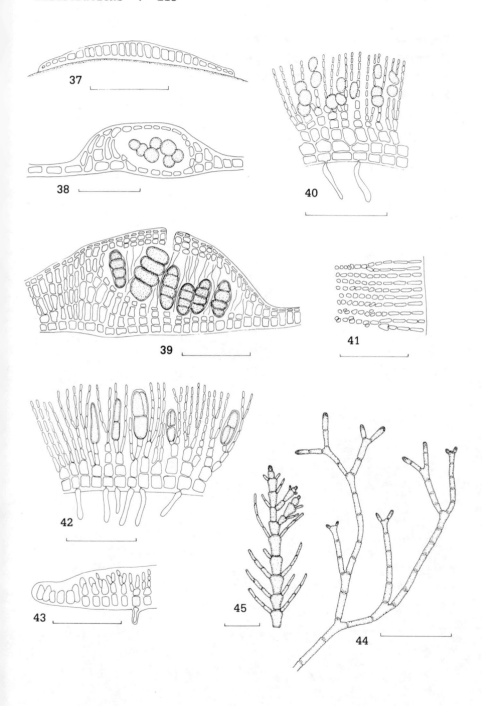

Figure 46. *Amphiroa beauvoisii*. Scale = 5 cm.
Figure 47. *Halymenia gelinaria*. Scale = 5 cm.
Figure 48. *Grateloupia filicina*.Scale = 3 cm.
Figure 49. *Halymenia floridana*. Scale = 3 cm.

Illustrations : 117

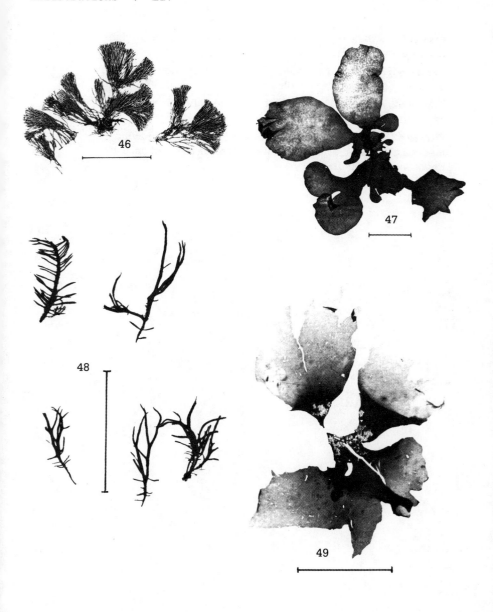

Figure 50. *Scinaia complanata*. Transection of an axis showing large polyhedral surface cells. Scale = 100 μm.

Figure 51. *Halymenia floridana*. Transection of blade showing medullary ganglia. Scale = 100 μm.

Figure 52. *Halymenia gelinaria*. Transection of blade showing segmented vertical filaments. Scale = 100 μm.

Figure 53. *Grateloupia filicina*. Transection of blade showing medulla of anastomosing filaments and cortex of anticlinal cell rows. Scale = 100 μm.

Illustrations : 119

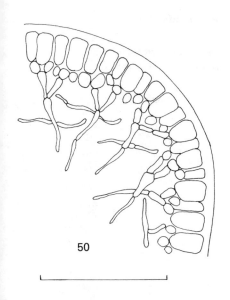

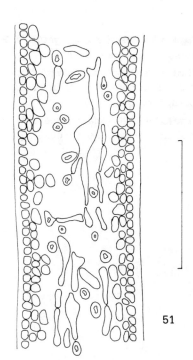

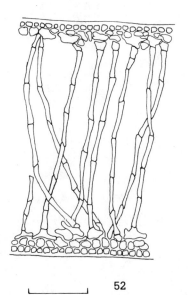

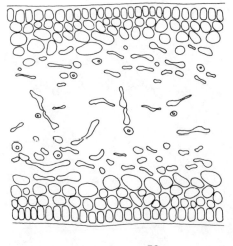

Figure 54. *Gracilaria blodgettii*. Note dense, spindlelike branchlets. Scale = 5 cm.

Figure 55. *Gracilaria foliifera*. Note flattened blades widest at points of branching. Some branches with epiphytic *Porphyra rosengurtii*. Scale = 5 cm.

Figure 56. *Gracilaria verrucosa*. Note terete branches ending in slender tips. Scale = 5 cm.

Illustrations : 121

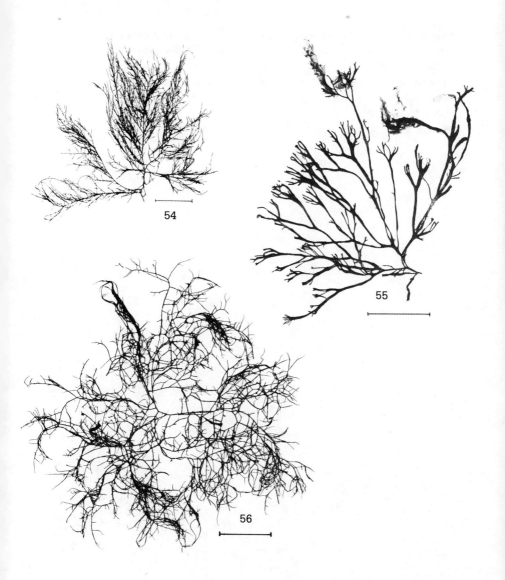

Figure 57. *Gracilaria verrucosa*. Transection of axis showing large, thin-walled medullary cells and cortex of 2 cell layers. Scale = 100 μm.

Figure 58. *Gracilaria foliifera*. Transection through cystocarpic region of axis. Note gonimoblast of relatively few large cells. Scale = 100 μm.

Figure 59. *Gracilaria sjoestedtii*. Cystocarpic plants. Scale = 5 cm.

Illustrations : 123

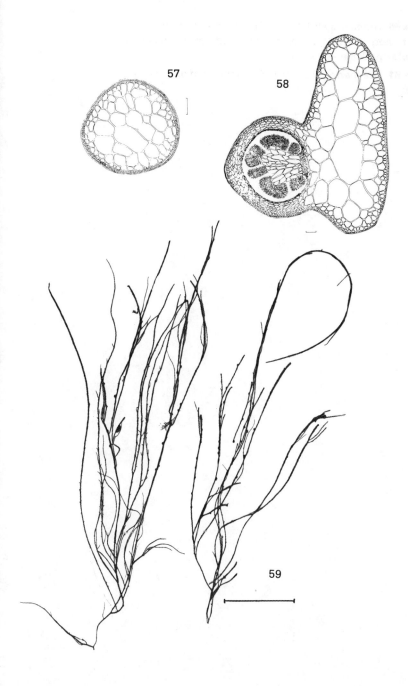

Figure 60. *Gracilaria sjoestedtii*. Transection through cystocarpic region of axis. Note highly filamentous, small-celled gonimoblast. Scale = 100 µm.

Figure 61. *Gracilaria blodgettii*. Transection of axis. Scale = 100 µm.

Illustrations : 125

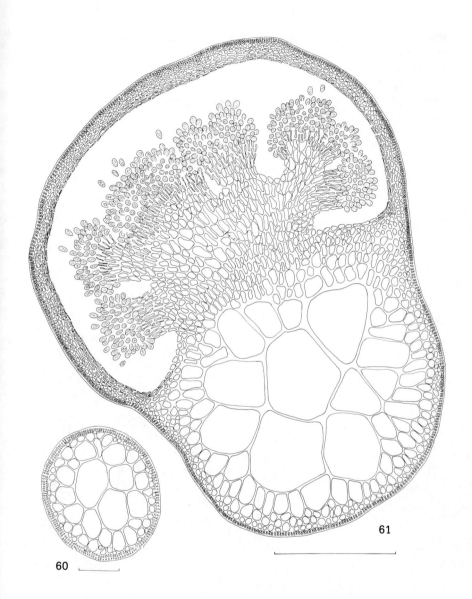

Figures 62-64. *Neoagardhiella baileyi*. Fig. 62. Cartilaginous form bearing cystocarps in papillae. Scale = 5 cm. Fig. 63. Typical soft form. Scale = 5 cm. Fig. 64. Flattened form with branching in one plane. Scale = 5 cm.

Figure 65. *Meristotheca floridana*. Note proliferous marginal branching. Scale = 5 cm.

Illustrations : 127

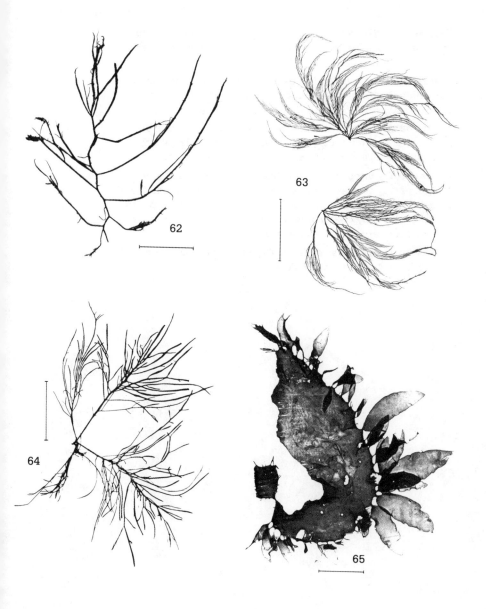

Figure 66. *Gigartina acicularis.* Transection of axis showing medulla of thick-walled cells and cortex of anticlinal cell rows. Scale = 100 µm.

Figure 67. *Gracilaria blodgettii.* Transection of spindlelike branchlet showing cruciate tetrasporangia in the cortex. Scale = 100 µm.

Figure 68. *Meristotheca floridana.* Transection of blade showing filamentous medullary cells connecting with large subcortical cells. Scale = 100 µm.

Figure 69. *Gymnogongrus griffithsiae.* Transection through margin of carpotetrasporophyte nemathecia bearing chains of cruciate tetraspores on the surface of a female gametophyte. In the gametophytic axis note medulla of thick-walled cells and cortex of anticlinal cell rows. Scale = 100 µm.

Figure 70. *Gracilaria sjoestedtii.* Transection of axis showing cortex of 3—4 cell layers. Scale = 100 µm.

Illustrations : 129

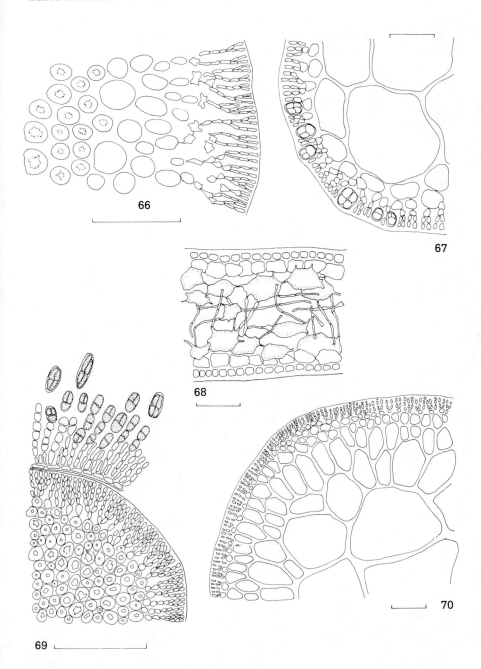

Figures 71-74. *Neoagardhiella baileyi*. Fig. 71. Transection of axis showing cortex of compact cells and filamentous inner medulla. Fig. 72. Transection of tetrasporangial region showing zonate tetrasporangia and loose cortex. Fig. 73. Transection of embedded cystocarp with parenchymatous gonimoblast tissue. Fig. 74. Transection of spermatangial region of cortex showing spermatia release. Scale = 100 µm.

Illustrations : 131

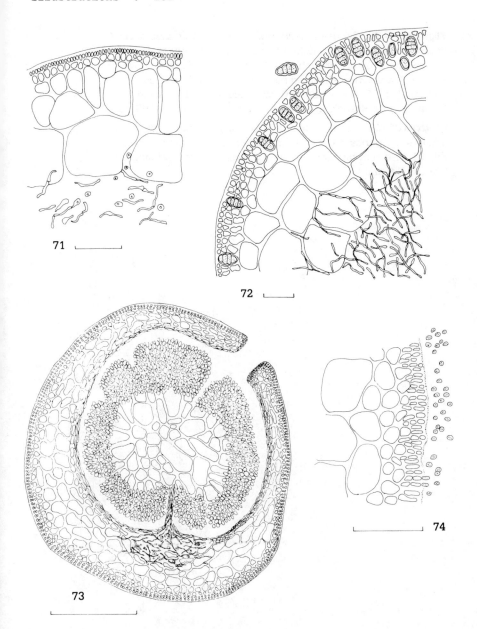

Figures 75-79. *Hypnea musciformis.* Fig. 75. Main axis beset with tetrasporangial branchlets swollen at their bases. Fig. 76. Hooked or "crozier" branch tips. Figs. 75—76. Scale = 1 mm. Fig. 77. Transection showing zonate tetrasporangia. Scale = 100 μm. Fig. 78. Main axis with spherical pericarps. Scale = 1 mm. Fig. 79. Branch tip showing apical cell. Scale = 100 μm.

Figures 80-81. *Hypnea cornuta.* Fig. 80. Main axis beset with urn-shaped pericarps and stellate branchlets. Scale = 1 mm. Fig. 81. Transection of axis showing persistent central filament. Scale = 0.5 mm.

Illustrations : 133

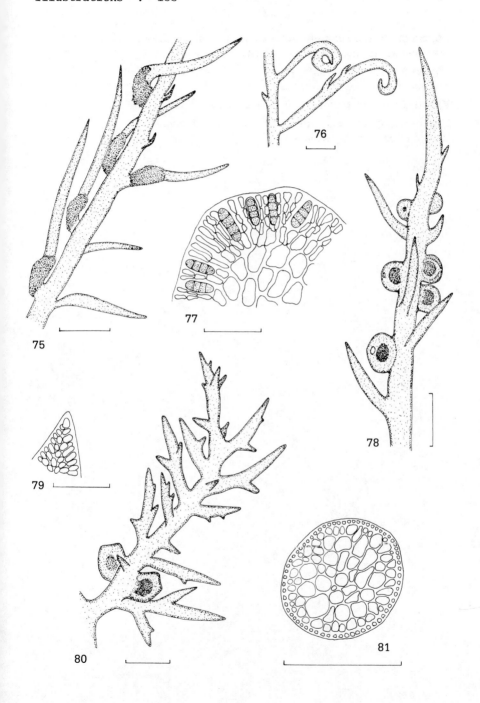

Figure 82. *Hypnea cornuta*. Note distinct main axis beset with stellate branchlets. Scale = 5 cm.

Figure 83. *Hypnea musciformis*. Note lax habit of alternately branched axes. Scale = 5 cm.

Figure 84. *Gymnogongrus griffithsiae*. Scale = 5 cm.

Figure 85. *Gigartina acicularis*. Note elongate branch tips devoid of branchlets. Scale = 5 cm.

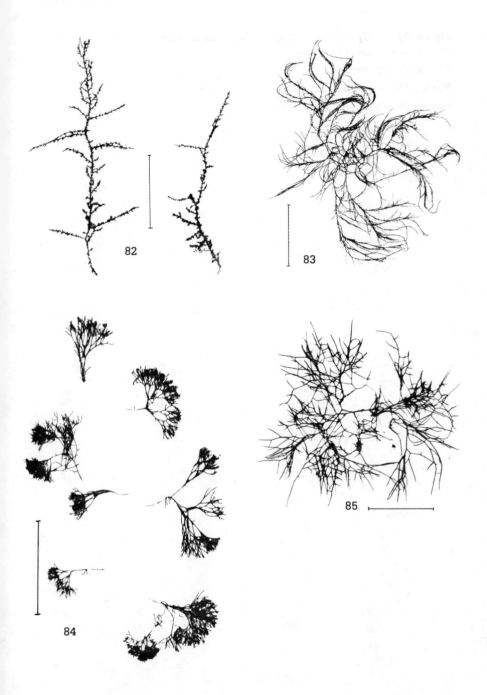

Figure 86. *Rhodymenia pseudopalmata.* Note spatulate blade tips with tetrasporangial sori. Scale = 5 cm.

Figure 87. *Botryocladia occidentalis.* Scale = 5 cm.

Figure 88. *Lomentaria baileyana.* Scale = 2 cm.

Figure 89. *Champia parvula.* Scale = 2 cm.

Illustrations : 137

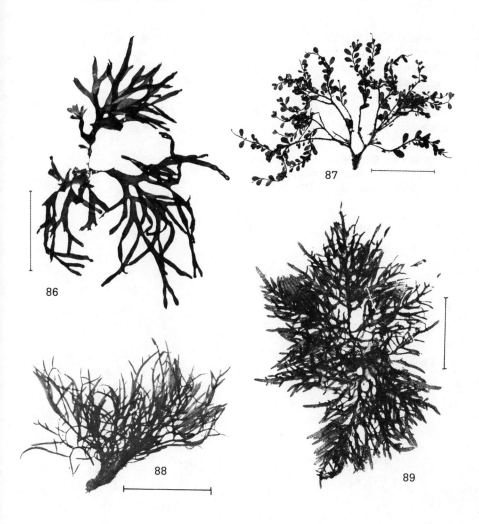

Illustrations : 138

Figures 90-92. *Champia parvula*. Fig. 90. Transection of cystocarpic region of axis showing hollow axis and carposporophyte with fertile gonimoblast filaments supporting carpospores, and a sterile mass of small-celled filaments. Scale = 100 μm. Fig. 91. Axis showing lateral pericarps and internal septa. Fig. 92. Axis showing pattern of alternate branching and superficial tetrasporangia. Figs. 91—92. Scale = 1 mm.

Figures 93-94. *Lomentaria baileyana*. Fig. 93. Axis with secund branches supporting pericarps. Fig. 94. Tetrasporangial branchlets. Figs. 93—94. Scale = 1 mm.

Illustrations : 139

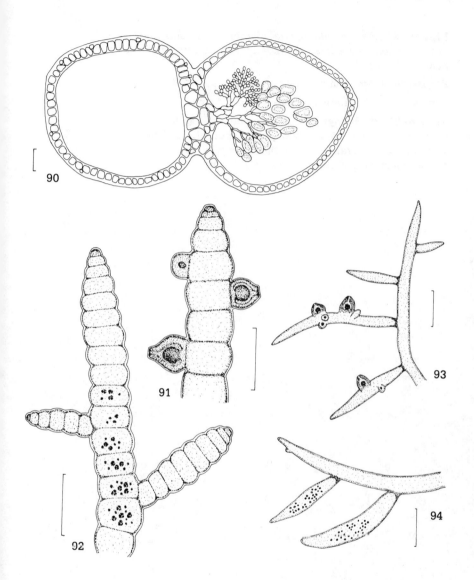

Figure 95. *Griffithsia globulifera*. Filament apex showing whorls of tetrasporangia subtended by blunt involucral cells. Scale = 100 µm.
Figure 96. *Spermothamnion investiens*. Scale = 100 µm.
Figure 97. *Griffithsia globulifera*. Scale = 1 mm.
Figures 98-99. *Rhodymenia pseudopalmata*. Fig. 98. Transection of blade margin showing carposporophyte. Scale = 100 µm. Fig. 99. Transection of blade showing cruciate tetrasporangia in cortical region. Scale = 200 µm.

Illustrations : 141

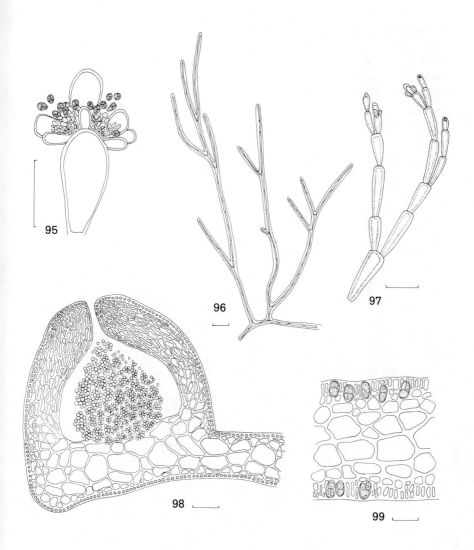

Figure 100. *Anotrichium tenue.* Scale = 100 µm.

Figures 101-102. *Anotrichium barbatum.* Plants collected epiphytic on pelagic *Sargassum fluitans*. Fig. 101. Pedicellate tetrasporangia. Fig. 102. Antheridial branches. Scale = 100 µm.

Figure 103. *Griffithsia globulifera.* Scale = 5 cm.

Figure 104. *Anotrichium tenue.* Scale = 3 cm.

Illustrations : 143

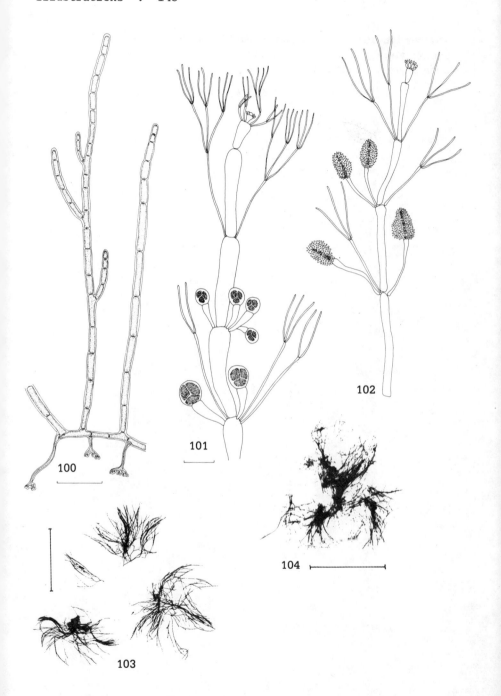

Figures 105-107. *Antithamnion cruciatum.* Fig. 105. Tetrasporophyte. Fig. 106. Antheridial branches. Fig. 107. Carposporophyte development. Scale = 100 µm.

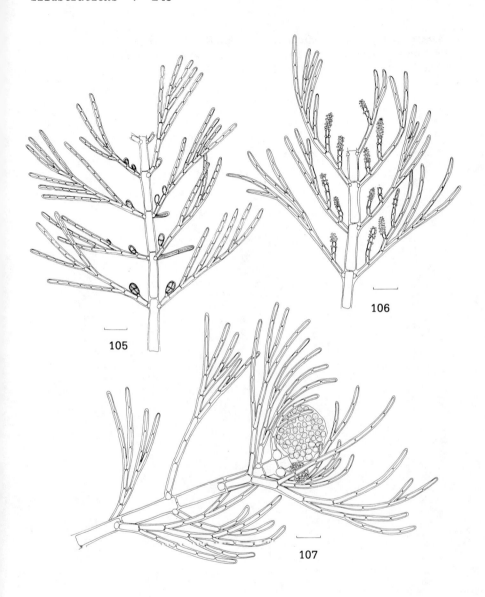

Figure 108-110. *Callithamnion byssoides*. Fig. 108. Carposporophyte development. Fig. 109. Tetrasporophyte. Fig. 110. Antheridial clusters. Scale = 100 μm.

Illustrations : 147

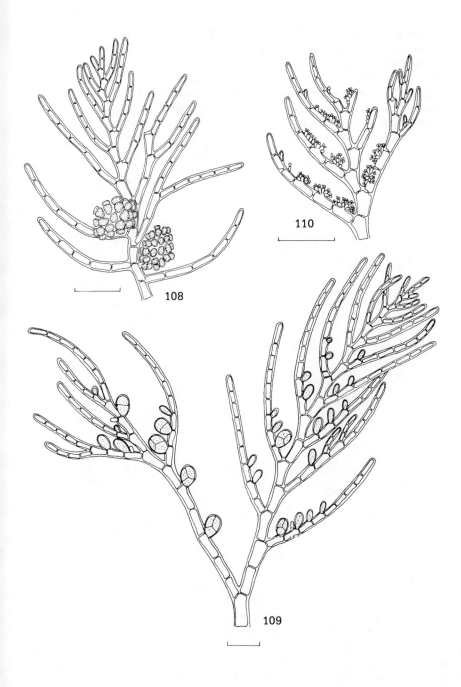

Figures 111-116. *Ceramium byssoideum.* Fig. 111, 115. Apex of tetrasporophytic branch. Note emergent tetrasporangia and horizontally elongated cells in the lower tiers of mature nodes. Fig. 112. Carposporophyte subtended by blunt branchlets. Fig. 113. Prostrate branch attached by unicellular rhizoids. Fig. 114. Vegetative nodal development with characteristic horizontally elongated lower cells. Fig. 116. Apex of branch showing deciduous hairs. Scale = 100 μm.

Illustrations : 149

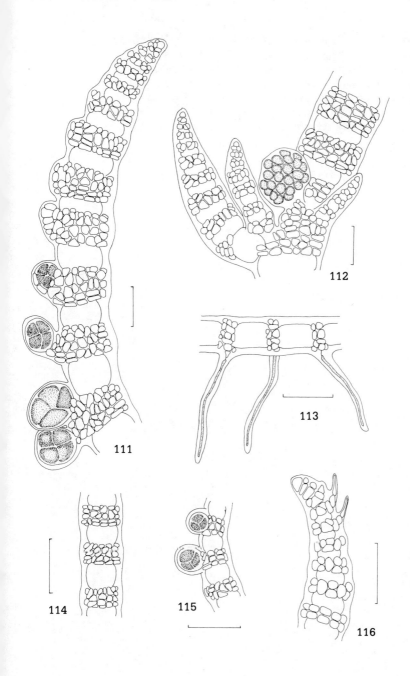

Figures 117-123. *Ceramium diaphanum*. Fig. 117. Nodal cortication of mature vegetative axis. Fig. 118. Apex with distinctive forcipate tips. Fig. 119. Axis showing adventitious branching and swollen deep-placed nodal cells. Figs. 120—121. Variation in tetrasporangial development. Fig. 112. Spermatia formation from cortical cells. Fig. 123. Carposporophyte without numerous subtending involucral branches. Scale = 100 μm.

Illustrations : 151

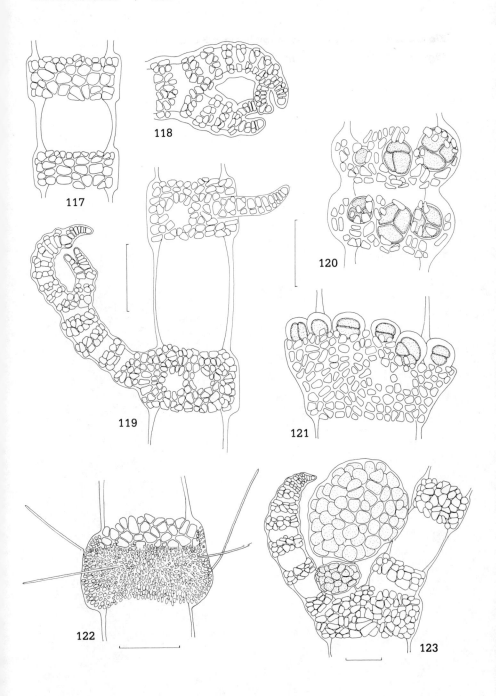

Figure 124. *Ceramium diaphanum*. Scale = 3 cm.
Figure 125. *Ceramium rubrum*. Scale = 5 cm.

Illustrations : 153

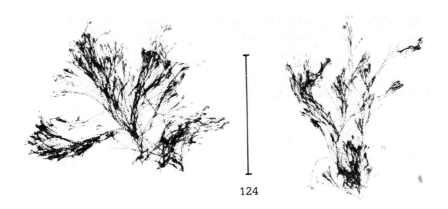

124

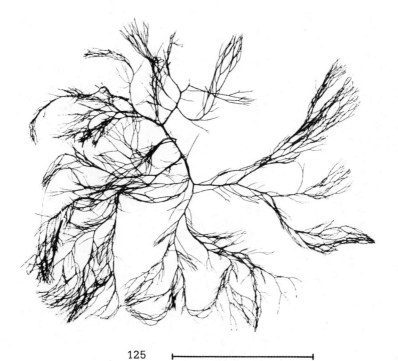

125

Figures 126-130. *Ceramium fastigiatum* f. *flaccidum*. Fig. 126. Bifurcate (but not forcipate) branch tips. Fig. 127. Carposporophytes borne in groups. Fig. 128. Emergent tetrasporangia from typical nodes of two-celled tiers. Fig. 129. Spermatia formation from cortical cells. Fig. 130. Prostrate branch attached by multicellular rhizoids. Scale = 100 µm.

Illustrations : 155

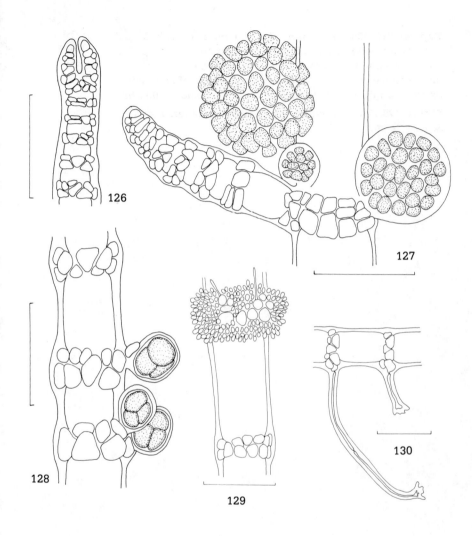

Figures 131-135. *Ceramium rubrum.* Figs. 131—132. Carposporophyte development. Fig. 133. Forcipate branch tip. Fig. 134. Tetrasporangial development in nodal bands and complete surface cortication. Fig. 135. Transection of node with tetrasporangia and continuous cortication. Scale = 100 µm.

Figure 136. *Spyridia hypnoides.* Note hooked branch tips. Scale = 5 cm.

Illustrations : 157

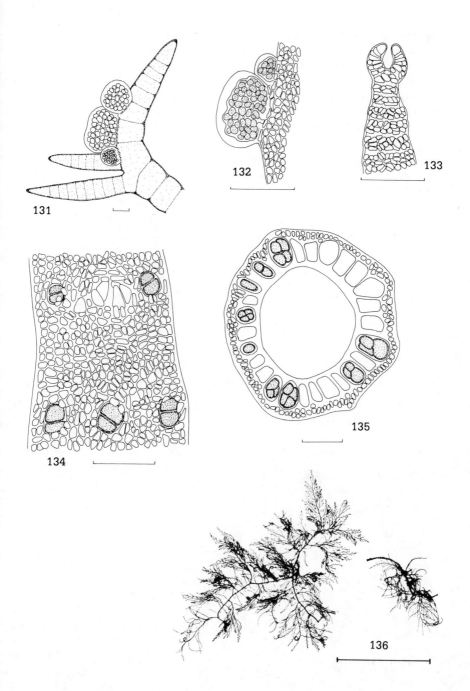

Figures 137-138. *Spyridia hypnoides.* Fig. 137. Variation in terminal spines on one plant. Fig. 138. Distinctive pattern of nodal and internodal cortication for this species. Scale = 100 μm.

Figures 139-140. *Micropeuce mucronata.* Fig. 139. Juvenile pericarp on lateral branch. Fig. 140. Main axis with corticating rhizoidal cells and monosiphonous ramelli. Scale = 100 μm.

Figures 141-142. *Dasya baillouviana.* Fig. 141. Main axis with complete superficial cortication and highly branched monosiphonous ramelli. Fig. 142. Tetrasporangial stichidia and associated ramelli. Scale = 100 μm.

Illustrations : 159

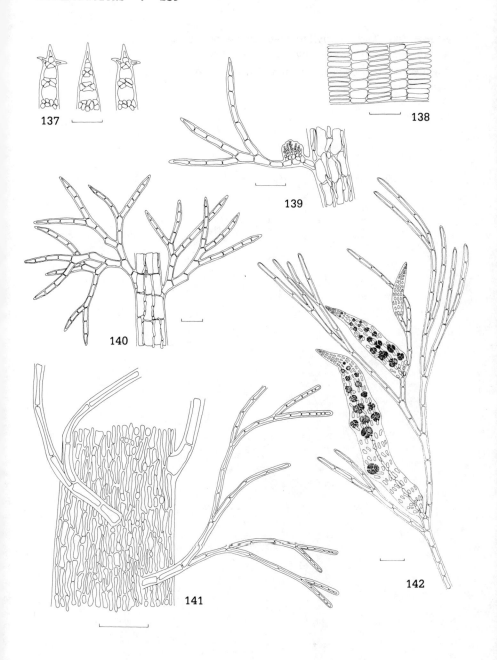

Figure 143. *Calonitophyllum medium.* Carposporophyte above, tetrasporophyte below. Scale = 5 cm.

Figure 144. *Grinnellia americana.* Scale = 5 cm.

Illustrations : 161

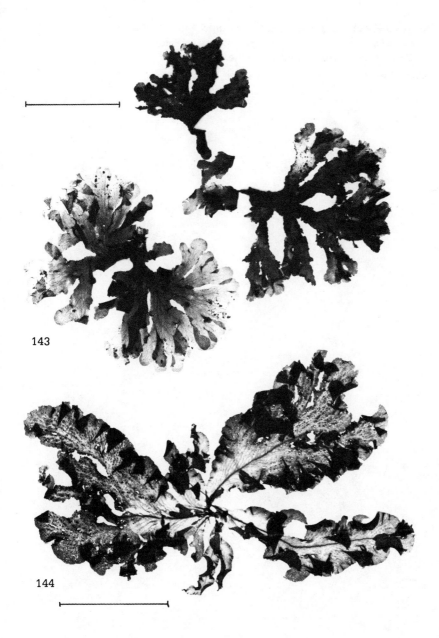

143

144

Figures 145-147. *Hypoglossum tenuifolium.* Fig. 145. Tetrasporangial sorus. Fig. 146. Carposporophyte on midrib. Fig. 147. Pattern of midrib branching. Scale = 1 mm.
Figure 148. *Branchioglossum minutum.* Scale = 1 mm.

Illustrations : 163

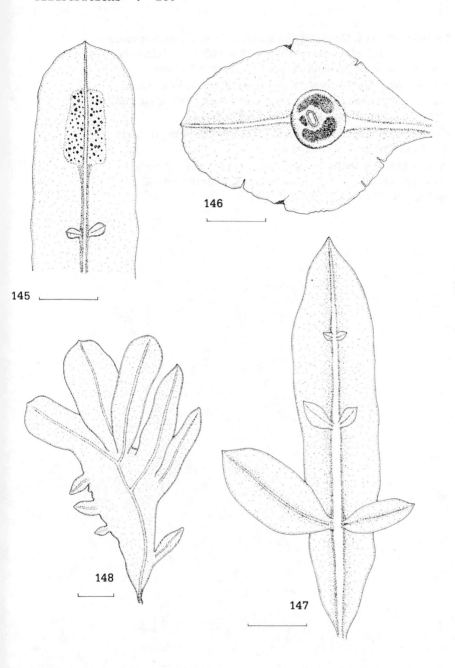

Figures 149-151. *Hypoglossum tenuifolium* var. *carolinianum*. Fig. 149. Tetrahedral tetrasporangia. Fig. 150. Branch initials from midrib. Fig. 151. Antheridial sori. Scale = 100 μm.

Figure 152. *Branchioglossum minutum*. Blade apex showing lens-shaped apical cell and conspicuous rows of elongate cells emanating from the midrib. Scale = 100 μm.

Figure 153. *Caloglossa lepriureii*. Blade apex showing dichotomous branching. Scale = 100 μm.

Figure 154. *Hypoglossum teniufolium*. Blade apex showing acute apical cell and polygonal cells in unordered rows emanating from the midrib. Scale = 100 μm.

Illustrations : 165

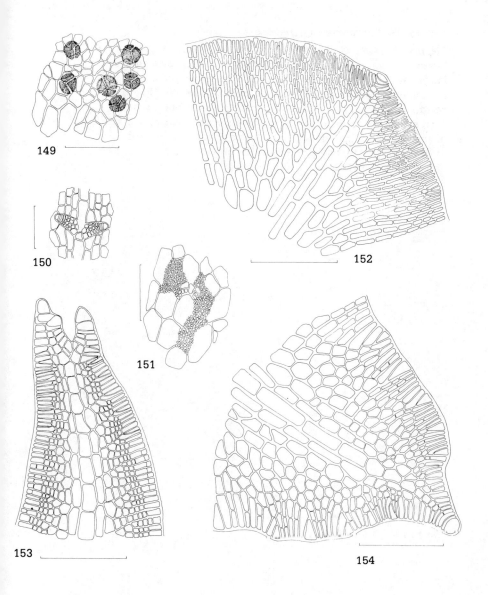

Figure 155. *Dasya baillouviana.* Scale = 5 cm.
Figures 156-161. *Chondria dasyphylla.* Fig. 156. Branchlet bearing pericarps. Fig. 157. Transection of mature axis showing parenchymatous appearance. Fig. 158. Tetrasporangial branchlets. Fig. 159. Transection of juvenile axis showing central cell with 5 pericentral cells giving rise to branches to form a cortical layer. Fig. 160. Branchlets with apical clusters of spermatangial platelets. Fig. 161. Spermatangial platelet with subtending trichoblast. Figs. 156, 158, 160. Scale = 1 mm. Figs. 157, 159, 161. Scale = 100 μm.

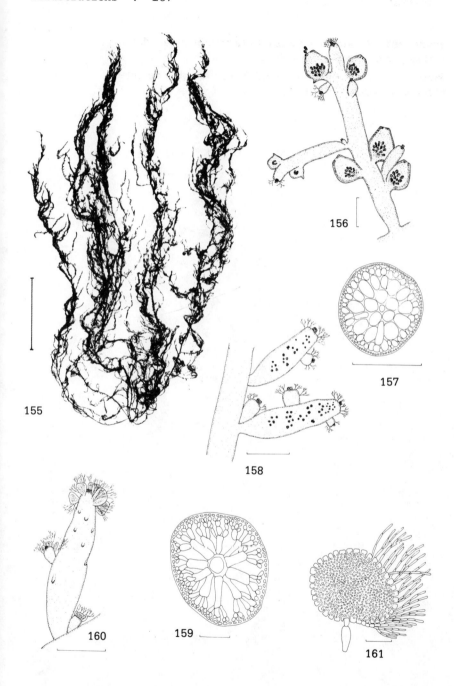

Figure 162. *Chondria dasyphylla.* Note broadly pyramidal outline of main axes. Scale = 5 cm.

Figure 163. *Micropeuce mucronata.* Note main axes denuded of branches. Scale = 5 cm.

Illustrations : 169

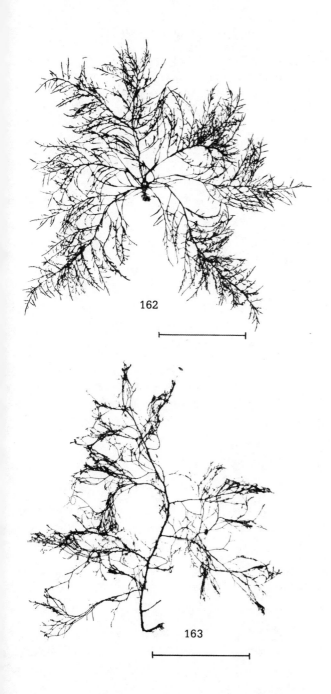

162

163

Figures 164-169. *Chondria littoralis.* Fig. 164. Branchlet with developing pericarps. Fig. 165. Transection of mature axis. Fig. 166. Tetrasporangial branchlet. Fig. 167. Branchlet with developing pericarp. Fig. 168. Branchlet with spermatangial platelets. Fig. 169. Tetrahedral tetrasporangia beneath cortical layer. Figs. 164, 166, 168. Scale = 1 mm. Figs. 165, 167, 169. Scale = 100 µm.

Figure 170. *Chondria littoralis.* Scale = 5 cm.

Illustrations : 171

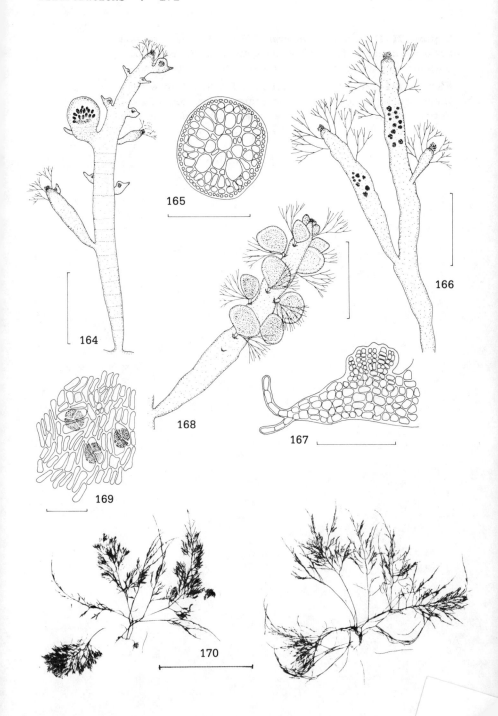

Figure 171-173. *Chondria polyrhiza*. Fig. 171. Creeping habit. Scale = 1 mm. Fig. 172. Rhizoidal cluster. Fig. 173. Branchlet with exposed apical cell. Figs. 172—173. Scale = 100 μm.

Figures 174-176. *Chondria tenuissima*. Fig. 174. Clusters of branchlets tapering to both ends. Scale = 1 mm. Fig. 175. Tetrasporangia. Fig. 176. Exposed apical cell. Scale = 100 μm.

Illustrations : 173

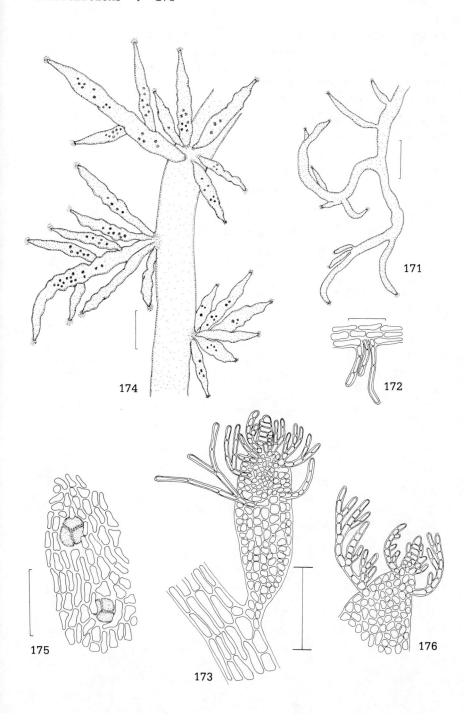

Figures 177-179. *Herposiphonia tenella.* Fig. 177. Antheridial branches borne in series along the upper branches. Fig. 178. Urceolate pericarp. Fig. 179. Development of erect branches and incurved apex. Scale = 100 µm.

Figure 180. *Laurencia poitei.* Scale = 5 cm.

Figure 181. *Laurencia corallopsis.* Note encrusting *Dermatolithon pustulatum.* Scale = 5 cm.

Illustrations : 175

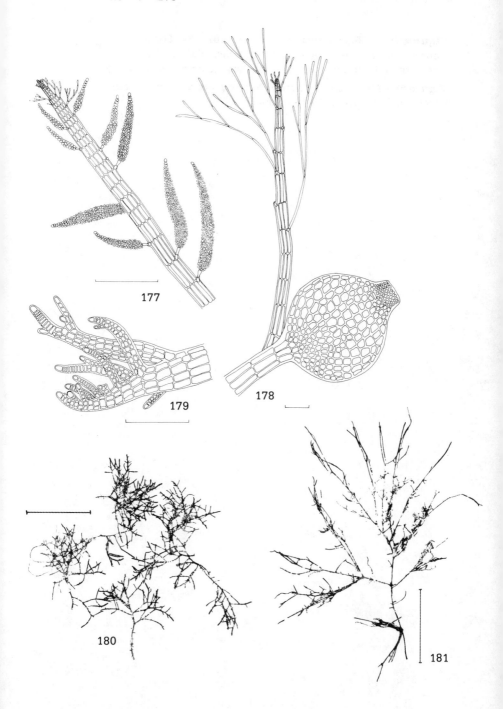

Figures 182-185. *Laurencia poitei*. Figs. 182, 184. Transections of mature axes. Scale = 0.5 mm. Figs. 183, 185. Branch tips with and without apical trichoblasts. Scale = 1 mm.

Figure 186. *Laurencia corallopsis*. Note square or radially elongated cortical cells. Scale = 0.5 mm.

Illustrations : 177

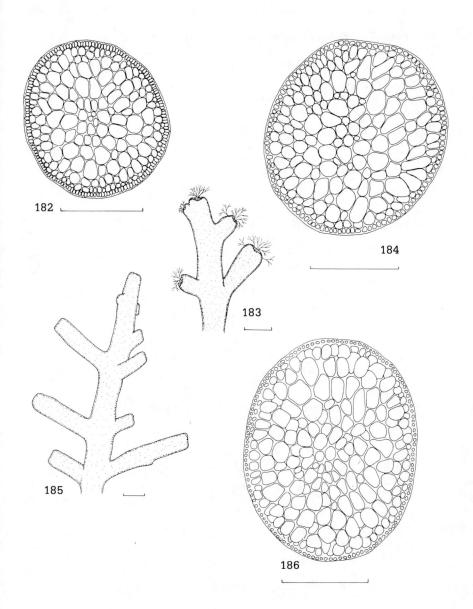

Figures 187-189. *Polysiphonia harveyi*. Fig. 187. Mature pericarps. Fig. 188. Cortication of main axis. Fig. 189. Branch arising to the side of a trichoblast. Scale = 100 μm.

Figures 190-192. *Polysiphonia sphaerocarpa*. Fig. 190. Mature pericarps. Fig. 191. Antheridial branches subtended by trichoblasts. Fig. 192. Prostrate branch attached by rhizoids cut off from pericentral cells. Scale = 100 μm.

Illustrations : 179

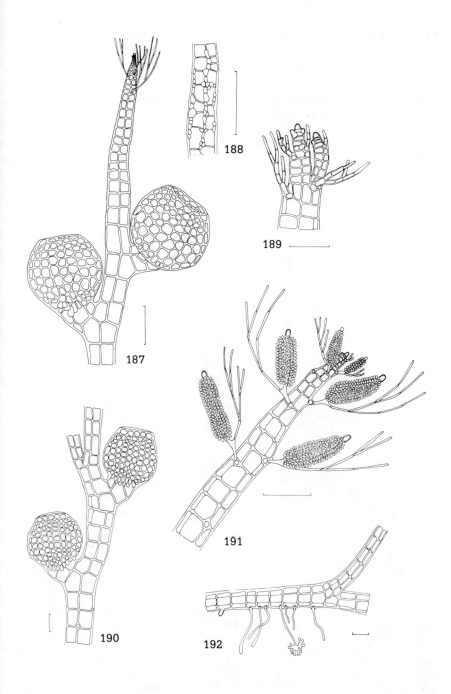

Figures 193-195. *Polysiphonia denudata.* Fig. 193. Antheridial branches terminated by 1—3 sterile cells. Fig. 194. Tetrasporangial branchlets. Fig. 195. Mature pericarp. Scale = 100 μm.

Figures 196-198. *Polysiphonia macrocarpa.* Fig. 196. Mature pericarp. Fig. 197. Prostrate axis attached by rhizoids remaining in open connection with pericentral cells. Fig. 198. Prostrate axis showing scar cells, endogenous and exogenous branching. Scale = 100 μm.

Illustrations : 181

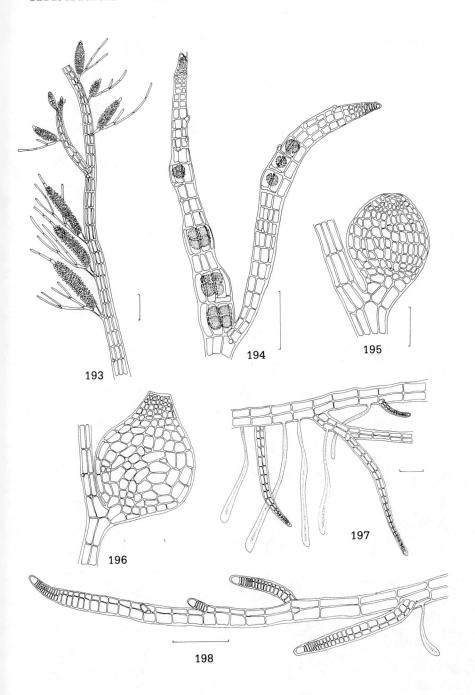

Figures 199-201. *Polysiphonia havanensis.* Fig. 199. Tetrasporangial branches. Fig. 200. Branch arising in axil of trichoblast. Fig. 201. Prostrate branch with rhizoids in open connection with pericentral cells. Scale = 100 µm.

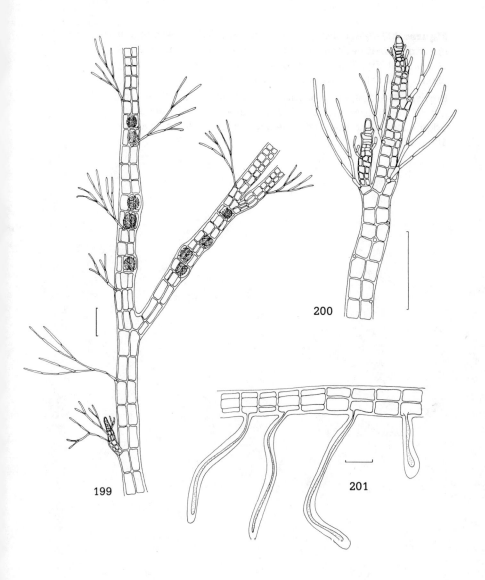

Figures 202-204. *Polysiphonia howei.* Fig. 202. Prostrate axis showing offset pericentral cells and rhizoid cut off from pericentral cell. Fig. 203. Apex with distinctive crest of trichoblasts. Fig. 204. Prostrate axis. Scale = 100 μm.

Figures 205-206. *Polysiphonia flaccidissima.* Fig. 205. Rhizoid cut off from pericentral cell. Fig. 206. Scar cells at branch origins indicating position of deciduous trichoblasts. Scale = 100 μm.

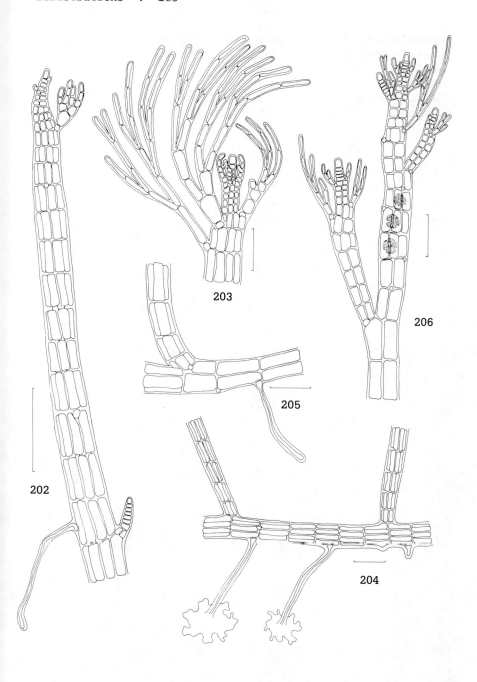

Figures 207-210. *Polysiphonia pseudovillum.* Epiphytic on pelagic *Sargassum fluitans.* Fig. 207. Forward bending erect branch. Fig. 208. Branch apex with trichoblasts. Fig. 209. Prostrate axis showing exogenous branch development and rhizoids cut off from pericentral cells. Fig. 210. Tetrasporangial branch. Scale = 100 µm.

Illustrations : 187

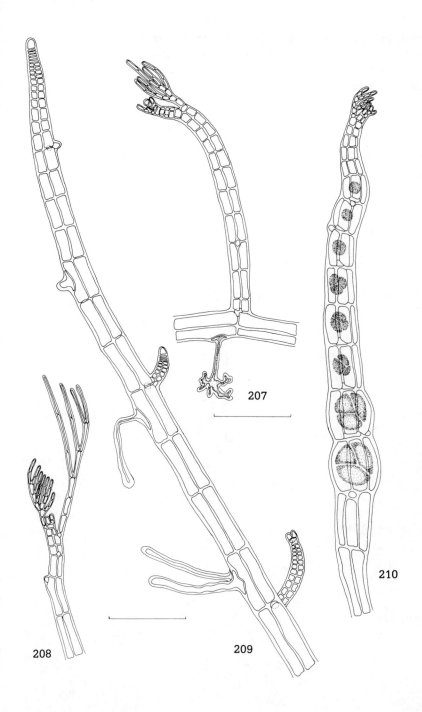

Figures 211-214. *Polysiphonia urceolata.* Fig. 211. Tetrasporangial branches. Fig. 212. Mature pericarp. Fig. 213—214. Spermatangial branches. Scale = 100 µm.
Figure 215. *Polysiphonia nigrescens.* Scale = 3 cm.

Illustrations : 189

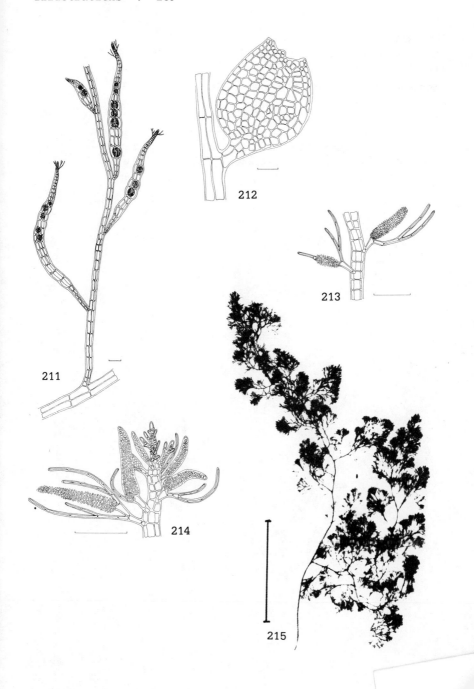

Figures 216-217. *Polysiphonia tepida.* Fig. 216. Bifurcate antheridial branches. Fig. 217. Branch arising in axil of trichoblast. Scale = 100 μm.

Figure 218. *Polysiphonia nigrescens.* Tetrasporangial branch. Scale = 100 μm.

Figures 219-221. *Polysiphonia ferulacea.* Fig. 219. Tetrasporangial branch. Fig. 220. Antheridial branch with sterile tip cells and subtending trichoblast. Fig. 221. Mature pericarps below and carpogonial branch with trichogyne above. Scale = 100 μm.

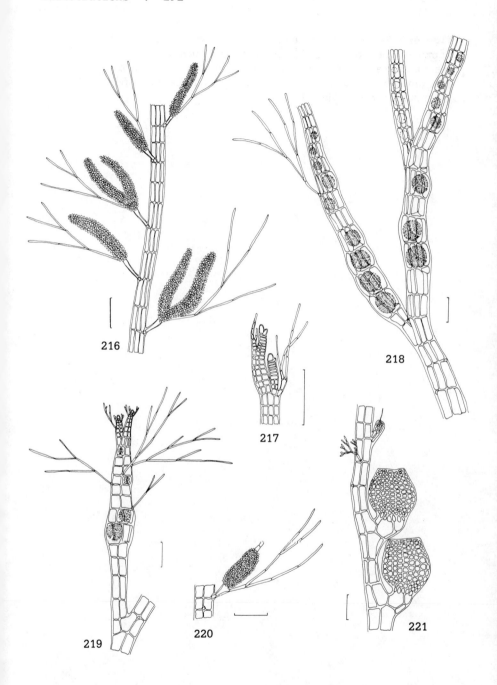

Figures 222-225. *Polysiphonia subtilissima.* Fig. 222. Erect apex with trichoblast. Fig. 223. Exogenous branch development from scar cell. Fig. 224. Rhizoids in open connection with pericentral cells. Fig. 225. Horizontal apex with radial development of branches. Scale = 100 µm.

Illustrations : 193

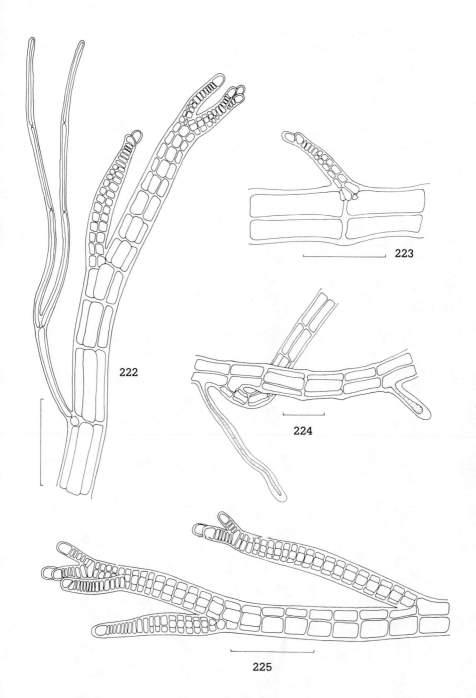

Figure 226. *Bostrychia radicans*. Monosiphonous terminal segments and stichidium. Scale = 100 μm.

Figure 227. *Pterosiphonia pennata*. Apex with incurved alternate branches. Scale = 100 μm.

Illustrations : 195

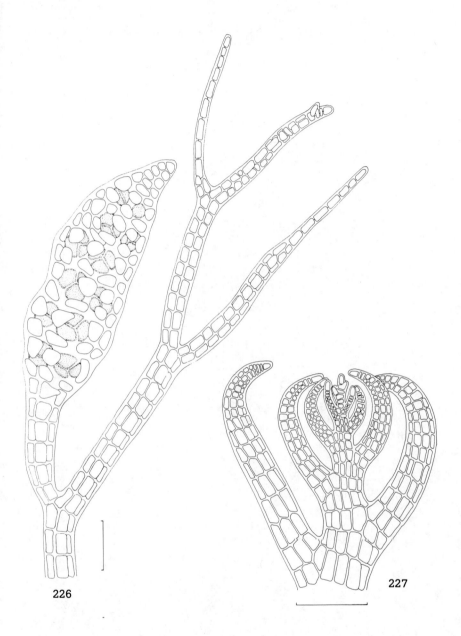

226 227

Glossary of Selected Terms

Adaxial: on the upper side of a branch.

Adventitious branching: branches arising secondarily below the apical primordia, usually from basal cells of deciduous trichoblasts.

Alternate branching: branches arising at different levels on opposite sides of an axis.

Anastomosing filaments: filaments joining together where individual cells come in contact, often forming ganglia.

Antheridia (spermatangia): region of a male gametophyte producing spermatia.

Arcuate branching: curved or crescent-shaped.

Benthic: growing attached, in contrast to pelagic (free-floating).

Cartilaginous: flexible, but not easily compressed.

Chromatophore (plastid): variously shaped organelle containing pigments.

Cicatrigenous: developing from a scar cell which is the remnant of a trichoblast.

Conceptacle: cavity containing reproductive structures and opening to the surface through an ostiole.

Cortex: small, secondarily derived cells external to larger medullary or pericentral cells.
Corticated: with an outer layer of small, secondarily derived cells covering larger apically derived cells.
Crozier tips: branch tips bent into a hook.
Crustose: thin and growing flattened against the substratum.
Cystocarp (carposporophyte): carpospore bearing tissue derived from a fertilized carpogonium and, usually, auxiliary cell tissue.
Denticulate margins: covered with microscopic teeth.
Dichotomous branching: branched by repeated forkings into two equal parts.
Discoidal base: disc-like pad of tissue which serves in some algae as an attachment structure.
Distromatic: consisting of only two cell layers.
Epilithic: growing attached to rocks.
Epiphytic: growing on other plants such as algae or seagrasses.
Eulittoral (littoral): region of the shore between high and low water marks.
Eurythermal: the ability to tolerate a wide range of temperatures.
Foliaceous: membranous or flattened like a leaf.
Forcipate: branch tips forked and incurved like tweezers.
Ganglia: cluster of interconnected (anastomosed) cells, typically in the medulla.
Gonimoblast: carpospore producing filaments in the cystocarp.
Haptera: branched structures for attachment to the substratum.
Hypothallus: the lower cell layer(s) of crustose algae. (The upper portion is the perithallus.)

Involucre: short, specialized branchlets arising beneath and partially covering reproductive structures.
Lanceolate: broadest at the base and narrowing toward the tip.
Littoral: see Eulittoral
Locule: chamber with reproductive structures.
Medulla: centermost vegetative cell region of an alga.
Monosiphonous: single row of cells without cortication, as in a branch tip.
Monosporangium: sporangium asexually producing single reproductive cells.
Monostromatic: having a single cell layer.
Nemathecium: a protruding chamber bearing reproductive structures in coralline algae.
Ostiole: opening of a pericarp or conceptacle.
Ovate: oval or egg-shaped.
Palmate: with marginal lobes arising from a central blade.
Parenchymatous: tissue consisting of isodiametric cells derived from cell divisions in more than one plane (not filamentous).
Papillae: small, wartlike projections.
Pericarp: sterile tissue covering the carposporophyte.
Pericentral cell: cell surrounding and cut off from a central filament or axial cell.
Pit connection: cytoplasmic strand connecting adjacent cells.
Plastid: see Chromatophore.
Plumose: featherlike with branchlets arranged in rows along opposite sides of an axis.
Pluriseriate: consisting of several cell rows.

Polysiphonous: composed of many pericentral cells (siphons) equal in length to the central axial cells which they surround.

Pseudoparenchymatous: tissue appearing parenchymatous, but structurally derived from compacted filaments.

Pulvinate: flattened pads or cushions.

Pyramidal: triangular, with the broadest dimension at the bottom.

Pyrenoid: organelle associated with starch storage in algal cells.

Ramelli: ultimate or most delicate branchlets.

Rhizine: filaments of thick-walled cells penetrating the medulla in members of the Gelidiales.

Rhizoids: attachment filaments of one to several nonpigmented cells.

Secund branching: unilateral; arranged in a row along one side of an axis.

Septate: partioned by a multicellular layer.

Sessile: attached directly to a branch; not stalked.

Setaceous: covered with fine, bristlelike filaments.

Stenothermal: the ability to tolerate only a narrow range of water temperatures.

Stichidium: a specialized tetraspore-bearing branch in red algae.

Subcortex: tissue layer immediately beneath the surface layer (cortex).

Sublittoral: region of the shore below low water mark.

Supralittoral: region of the shore above high water mark (splash zone).

Terete: cylindrical.

Tetrasporangium: sporangium containing four spores produced by one of three cleavage patterns: zonate or linear, cruciate or cross-shaped, and tetrahedral.
Thallus: botanical term for plant body lacking vascular tissue.
Trichoblast: uniseriate branched or simple filament of colorless cells derived from apical cell division.
Trichogyne: elongated tip of the carpogonium which receives the spermatia.
Uniseriate: consisting of a single row of cells.
Urceolate: urn-shaped.

Index

Acrochaetiaceae 12, 38
Amphiroa beauvoisii 12, 47; Fig. 46
Anotrichium barbatum 9, 14, 59; Figs. 101-102
Anotrichium tenue 14, 59; Figs. 100, 104
Antithamnion cruciatum 14, 60; Figs. 105-107
Audouinella alariae 12, 40; Figs. 15-17
Audouinella dasyae 12, 40; Figs. 13-14
Audouinella hallandica 12, 41; Figs. 18-20, 26
Audouinella microscopica 12, 41; Figs. 21-22
Audouinella secundata 12, 41; Figs. 23-25
Audouinella thuretii 12, 41; Fig. 27

Bangia atropurpurea 11, 35; Fig. 10
Bangiaceae 11, 35
Bangiales 11, 33
Bangiophyceae 11, 33
Bostrychia radicans 15, 70; Fig. 226
Botryocladia occidentalis 13, 57; Fig. 87
Branchioglossum minutum 7, 9, 14, 66; Fig. 148

Callithamnion byssoides 7, 14, 61; Figs. 108-110
Caloglossa leprieurii 14, 67; Fig. 153
Calonitophyllum medium 14, 67; Fig. 143
Ceramiaceae 14, 59
Ceramiales 14, 59
Ceramium byssoideum 14, 61; Figs. 111-116

Ceramium diaphanum   14, 62; Figs. 117-124
Ceramium fastigiatum   14, 63; Figs. 126-130
Ceramium rubrum   14, 63; Figs. 125, 131-135
Chaetangiaceae   12, 42
Champia parvula   14, 57; Figs. 90-92
Champiaceae   14, 57
Chondria dasyphylla   15, 71; Figs. 156-162
Chondria littoralis   15, 71; Figs. 164-170
Chondria polyrhiza   15, 72; Figs. 171-173
Chondria tenuissima   15, 72; Figs. 174-176
Corallina cubensis   13, 48; Fig. 45
Corallinaceae   12, 46
Cryptonemiales   12, 45

Dasya baillouviana   15, 69; Figs. 141-142, 155
Dasyaceae   15, 69
Delesseriaceae   14, 66
Dermatolithon pustulatum   12, 47; Fig. 39

Erythrocladia subintegra   11, 34; Fig. 2
Erythropeltidaceae   11, 33
Erythrotrichia carnea   11, 34; Figs. 6-8
Erythrotrichia ciliaris   11, 34; Fig. 4

Florideophyceae   11, 38

Gelidiales   12, 42
Gelidiaceae   12, 42
Gelidium americanum   12, 45; Figs. 29, 34-35
Gelidium crinale   12, 44; Figs. 30-33, 36
Gigartina acicularis   13, 56; Figs. 66, 85
Gigartinaceae   13, 56

Gigartinales 13, 50

Goniotrichales 11, 33

Goniotrichaceae 11, 33

Goniotrichum alsidii 11, 33; Fig. 3

Gracilaria blodgettii 9, 13, 50; Figs. 54, 61, 67

Gracilaria foliifera 13, 50; Figs. 55, 58

Gracilaria sjoestedtii 13, 51; Figs. 59-60, 70

Gracilaria verrucosa 13, 52; Figs. 56-57

Gracilariaceae 13, 50

Grateloupia filicina 13, 48; Figs. 48, 53

Grateloupiaceae 13, 48

Griffithsia globulifera 14, 64; Figs. 95, 97, 103

Grinnellia americana 14, 68; Fig. 144

Gymnogongrus griffithsiae 13, 55; Figs. 69, 84

Halymenia floridana 9, 13, 49; Figs. 49, 51

Halymenia gelinaria 13, 49; Figs. 47, 52

Herposiphonia tenella 15, 73; Figs. 177-179

Heteroderma lejolisii 12, 47; Figs. 37-38

Hypnea cornuta 13, 55; Figs. 80-82

Hypnea musciformis 13, 55; Figs. 75-79, 83

Hypneaceae 13, 54

Hypoglossum tenuifolium 14, 69; Figs. 145-147, 149-151, 154

Jania adhaerens 13, 48; Fig. 44

Laurencia corallopsis 15, 74; Figs. 181, 186

Laurencia poitei 15, 74; Figs. 180, 182-185

Lomentaria baileyana 14, 58; Figs. 88, 93-94

Meristotheca floridana 13, 53; Figs. 65, 68

Micropeuce mucronata 15, 75; Figs. 139-140, 163

Index : 206

Nemaliales 11, 38

Neoagardhiella baileyi 13, 53; Figs. 62-64, 71-74

Peyssonnelia rubra 12, 45; Figs. 40-43

Phyllophoraceae 13, 55

Polysiphonia denudata 15, 76; Figs. 193-195

Polysiphonia ferulacea 15, 76; Figs. 219-221

Polysiphonia flaccidissima 9, 15, 77; Figs. 205-206

Polysiphonia harveyi 15, 78; Figs. 187-189

Polysiphonia havanensis 15, 78; Figs. 199-201

Polysiphonia howei 9, 15, 79; Figs. 202-204

Polysiphonia macrocarpa 15, 80; Figs. 196-198

Polysiphonia nigrescens 15, 80; Figs. 215, 218

Polysiphonia pseudovillum 9, 15, 81; Figs. 207-210

Polysiphonia sphaerocarpa 15, 81; Figs. 190-192

Polysiphonia subtilissima 15, 82; Figs. 222-225

Polysiphonia tepida 15, 83; Figs. 216-217

Polysiphonia urceolata 15, 84; Figs. 211-214

Porphyra carolinensis 7, 11, 36; Figs. 9, 11

Porphyra rosengurtii 7, 11, 37; Figs. 5, 12

Pterosiphonia pennata 15, 84; Fig. 227

Rhodomelaceae 15, 70

Rhodymenia pseudopalmata 13, 56; Figs. 86, 98-99

Rhodymeniaceae 13, 56

Rhodymeniales 13, 56

Scinaia complanata 12, 42; Figs. 28, 50

Solieriaceae 13, 52

Spermothamnion investiens 14, 64; Fig. 96

Spyridia hypnoides 14, 65; Figs. 137-138

Squamariaceae 12, 45

THE LIBRARY
ST. MARY'S COLLEGE OF MARYLAND
ST. MARY'S CITY, MARYLAND 20686